Lecture Notes in Mathematics Vol. 1673
ISBN 978-3-540-63628-1 © Springer-Verl

Frank D. Grosshans

Algebraic Homogeneous Spaces and Invariant Theory

Errata

My thanks to Nazih Nahlus, Walter Ferrer Santos, and especially Michel Brion for the errata listed below.

page 8, line −7: "$T(G(x))_x$" should be $T(G \cdot x)_x$.

page 13, line −7: "we adapt" should be "we adopt".

page 15, line 15: "(ω, α)". The notation $(\ ,\ )$ refers to a $\mathbb{Q}$ valued inner product on $X(T)$ which is invariant under the Weyl group of G.

page 20, statement of Theorem 4.2: "$k[X'] = k[X]$ if and only if $\operatorname{codim}(X - X') \geq 2$". This would be better stated as "$k[X'] = k[X]$ if and only if each irreducible component of $X - X'$ has codimension ≥ 2 in X".

page 27, line −2: "dimensions of the irreducible components of the fibers" should be "dimensions of the irreducible components of the non-empty fibers".

page 28, Theorem 5.3: the assumption that X be affine is not necessary.

page 29, line −4: The sentence, "Embeddings of G/U, where U is maximal unipotent in a semi-simple group G, were completely described in [112] in case char $k = 0$", should be replaced by the following. "The affine embeddings of G/U, where U is maximal unipotent in a semi-simple group G, were completely described in [112] in case char $k = 0$. A general theory of embeddings of homogeneous spaces was developed by Luna and Vust in [70]. An exposition of the Luna-Vust theory may be found in [60]."

page 39, line 18: the bibliographic citation should be [Ferrer Santos, Walter Ricardo, A note on affine quotients. J. London Math. Soc. (2) 31 (1985), no. 2, 292–294].

page 44, line −1: "there is non-negative" should be "there is a non-negative".

page 49, line 5: "Lemma 3 is, in fact" should be "Lemma 8.5 is, in fact".

page 50, line −6: "Then $\Phi(a) \neq 0$" should be "If $a \neq 0$, then $\Phi(a) \neq 0$".

page 59, line −13: "a subgroup H" should be "an observable subgroup H".

page 63, line 1: "$SL(n, \mathbb{C})$" should be "$SL(n, \mathbb{C}), n \geq 2$".

page 66, line 7: "the group acts on S_n" should be "the group G acts on S_n".

page 73, line 10. The sentence beginning "Since $s_o\omega \leq \chi \leq \omega$" should be replaced by the following. "Since $\omega \geq \chi$ and $\chi + \sum e_\alpha \alpha = s_o\omega$, we have $\sum e_\alpha \alpha = s_o\omega - \chi \geq s_o\omega - \omega$. Hence, $s_o\omega - \omega \leq \sum e_\alpha \alpha \leq 0$."

page 93, proof of Theorem 15.14. Michel Brion has shown me an easy way to prove that D is a free $k[x]$-module. Since free modules are flat, this eliminates the need to use Lemma 15.9. The proof is as follows. For each $n \geq 0$, choose a subspace V_n of A_n so that as vector spaces over k, $A_n = V_n \oplus A_{n-1}$; choose a basis for V_n over k say, $\{v_{n1}, v_{n2}, \ldots\}$. Let $F = \{v_{01}, \ldots, v_{11}, \ldots\}$. Note that any $a \in A_n$ can be written as a k-linear combination of the v_{ij}. We now show that D is a free $k[x]$-module with F as a basis. First, we prove by induction on degree that any $f \in D$ is a $k[x]$-linear combination of elements in F. When $f \in A_0$, this is immediate since $A_{-1} = \{0\}$. In general, let $f = \sum_{n=0}^{N} a_n x^n$ where a_N is a non-zero element in A_N. Choose (finitely many) scalars $c_{ij} \in k$ so that $a_N = \sum c_{ij} v_{ij}$. Then, $f - \sum c_{ij} v_{ij} x^N \in \bigoplus_{n=0}^{N-1} A_n x^n$ and we may apply the induction hypothesis. Second, suppose that $\sum c_{ij}(x)\, v_{ij} = 0$ where each $c_{ij}(x)$ is a polynomial in $k[x]$. We equate the coefficient of each power of x to 0 to obtain equations of the form $\sum d_{ij}\, v_{ij} = 0$ where each $d_{ij} \in k$. Since the v_{ij} are linearly independent over k, each $d_{ij} = 0$. Then, each $c_{ij}(x) = 0$.

page 105, line 3: "$D(m, n, r)$" should be "$k[D(m, n, r)]$".

page 107, line −5. The following comment should be placed right after the definition. "This definition makes sense when $G = B$. Lemmas 19.7, 19.8, and Theorem 20.2 also are valid in case $G = B$ and are used in that way in the Example on p. 116."

page 108, Lemma 19.8: the lemma holds without the assumption that X be quasi-affine [Knop, Friedrich: On the set of orbits for a Borel subgroup. Comment. Math. Helv. 70 (1995), no. 2, 285–309].

page 117, line 13: "commutator" should be "centralizer".

page 121, line −19: "fourteen different" should be "thirteen different".

Lecture Notes in Mathematics 1673

Springer
Berlin
Heidelberg
New York
Barcelona
Budapest
Hong Kong
London
Milan
Paris
Santa Clara
Singapore
Tokyo

Frank D. Grosshans

Algebraic Homogeneous Spaces and Invariant Theory

Springer

Author

Frank D. Grosshans
Department of Mathematics
Wester Chester University of Pennsylvania
West Chester, PA 19383, USA
E-mail: fgrosshans@wcupa.edu

Cataloging-in-Publication Data applied for

Die Deutsche Bibliothek - CIP-Einheitsaufnahme

Grosshans, Frank D.:
Algebraic homogeneous spaces and invariant theory / Frank D. Grosshans. - Berlin ; Heidelberg ; New York ; Barcelona ; Budapest ; Hong Kong ; London ; Milan ; Paris ; Santa Clara ; Singapore ; Tokyo : Springer, 1997
(Lecture notes in mathematics ; 1673)
ISBN 3-540-63628-5

Mathematics Subject Classification (1991): 13A50, 14L30, 15A72, 20G15

ISSN 0075-8434
ISBN 3-540-63628-5 Springer-Verlag Berlin Heidelberg New York

Typesetting: Camera-ready TEX output by the author
SPIN: 10553372 46/3142-543210 - Printed on acid-free paper

Contents

Introduction

In these notes, we shall be concerned with the relationship between algebraic homogeneous spaces and certain questions in invariant theory. To be more precise, we need some definitions. Let k be an algebraically closed field of arbitrary characteristic. We shall assume that all varieties are defined over k and will always identify a variety with its k-rational points. Let A be a commutative k-algebra with multiplicative identity and let G be a linear algebraic group having identity e. A *rational action of G on A* is given by a mapping $G \times A \to A$, denoted by $(g,a) \to g{\cdot}a$ (or sometimes simply by ga), such that the following conditions are satisfied: (i) $(gg'){\cdot}a = g{\cdot}(g'{\cdot}a)$ and $e{\cdot}a = a$ for all $g,g' \in G$ and $a \in A$; (ii) the mapping $a \to g{\cdot}a$ is a k-algebra automorphism of A for all $g \in G$; (iii) every element of A is contained in a finite-dimensional subspace of A which is invariant under G and on which G acts by a rational representation. For H a closed subgroup of G, we put $A^H = \{a \in A : h{\cdot}a = a \text{ for all } h \in H\}$. Then, A^H is a k-algebra and this algebra is our main object of study, especially when H is not reductive.

Now we can describe some of the main themes which run throughout this book. We should begin by noting that most of our results are proved for fields of arbitrary characteristic and for groups which need not be connected. (The connected component of the identity in G will be denoted by G°.)

Observable subgroups.

A closed subgroup H of G is *observable* if G/H is an open subset of some affine variety. We give a number of conditions which are equivalent to this including one which is purely group-theoretic (§1-3, §7). Observable subgroups are of interest in representation theory since any finite-dimensional representation of an observable subgroup H may be extended to a finite-dimensional representation of G.

The algebra *k[G/H]*.

In Chapter One, we show that in studying $k[G/H]$, we may assume that H is observable in G. In this context, we show that $k[G/H]$ is finitely generated over k if and only if G/H can be embedded as an open subset of an affine variety X in such a way that $\dim(X - G/H) \leq \dim X - 2$ (§4). An important instance of this occurs when H is a maximal unipotent subgroup of G (§5).

The adjunction argument.

The adjunction argument states that $A^H = (k[G/H] \otimes_k A)^G$ where G acts on $k[G/H]$ by left translation. The idea for studying the invariants of H on A in terms of the invariants of the larger group G on a larger set, namely $(k[G/H] \otimes_k A)$, was discovered and used extensively in the nineteenth century. We look at several examples in geometry and physics of great historic interest (§10, 11). If G is reductive and if $k[G/H]$ and A are finitely generated over k, then the adjunction argument may be used to show that A^H is also finitely generated over k (§9). It is in this way that we obtain most of our results on the finite generation of rings of invariants of non-reductive groups.

Induced representations.

Let V be a finite-dimensional H-module. The action of H on V together with the action of H by right translation on $k[G]$ give an action of H on $k[G] \otimes_k V$. The action of G by left translation on $k[G]$ gives an action of G on $k[G] \otimes_k V$. Since the actions of H and G commute, G acts on $(k[G] \otimes_k V)^H$. This action is called the *induced module of V from H to G* (§6). Induced modules will be used to study observable subgroups (§7), prove the adjunction argument (§6), and find examples where the algebra of invariants is not finitely generated (§8).

Now, suppose that $k[G/H]$ is finitely generated over k and that V is finite-dimensional; we shall also consider whether the induced module is finitely generated over $k[G/H]$. For observable H, it is relatively easy to show finite generation, but the situation for subgroups which are not observable is not so clear (§23).

Maximal unipotent subgroups.

Of all the subgroups which we study, the most important are the maximal unipotent subgroups of reductive groups. Let G be reductive and let $B = TU$ be a Borel subgroup (of G°) where T is a maximal torus and U is a maximal unipotent subgroup. The group G acts by left translation on $k[G/U]$ and we study the structure of this representation in some detail using the general linear group as an example (§12,13).

In §15, we construct a graded algebra, gr$A$, which plays an important role in much of what follows; roughly speaking, this algebra often provides the machinery for proving results in arbitrary characteristic. For example, we use it to show that A is finitely generated over k if and only if A^U is finitely generated over k (§16) along with an analogous statement for normality (§18). For A-modules M, we show that M is finitely generated over A if and only if M^U is finitely generated over A^U. In particular, this means that a rational G-module V has finite dimension if and only if V^U does.

Subgroups of G which contain a maximal unipotent subgroup are called *horospherical*; their study, especially with respect to affine embeddings of G/H, is the object of §17.

We are also frequently interested in the actual construction of U-invariants. In a number of instances - including certain $k[G/H]$, binary forms, determinantal varieties, and the coadjoint representation - we show how to construct the algebra A^U.

Complexity.

Let B be a Borel subgroup of G, and let G act on an irreducible variety X. For χ a character of B, we define $k[X]_\chi = \{f \in k[X] : b{\cdot}f = \chi(b)f \text{ for all } b \in B\}$. The *complexity* of the action of G on X, $c(X)$, is the codimension in X of the B-orbit having highest dimension (§19). If $c(X) = 0$, the action of G on $k[X]$ is *multiplicity-free*, i.e, dim $k[X]_\chi \leq 1$ for all χ; such actions lie behind many of the explicit examples which we give (§11). A *spherical subgroup H of G* is one such that $c(G/H) = 0$; some basic properties are proved in §22. If $c(X) = 1$, precise information is also available on dim

$k[X]_X$ (§20). In the case where G is reductive and $c(X) \leq 1$ (along with other assumptions on $k[X]$), we show that $k[X]$ is finitely generated over k (§21). However, the famous Nagata counter-example to Hilbert's fourteenth problem gives an instance of a variety X where $c(X) = 2$ and $k[X]$ is not finitely generated over k.

Reductive groups.

First, a word about terminology is in order. We define the *radical of G*, $\mathfrak{R}_u G$, to be the radical of G° as defined in [10; 11.21]. Then, G is called reductive if its radical is $\{e\}$, i.e., if G° is reductive in the sense of [10; 11.21]. The invariant theory of reductive groups is of great importance in these notes. In particular, we shall assume the following theorem.

Theorem A. *Let G be a reductive group.*
(a) *Let A be a finitely generated, commutative k-algebra and let G act rationally on A. Then the algebra A^G is finitely generated over k.*
(b) *Let X be an affine variety on which G acts morphically and let Y be the affine variety with $k[Y] = k[X]^G$. Let $\pi: X \to Y$ be the mapping corresponding to the inclusion of k[Y] in k[X]. Then, (i) π is surjective; (ii) if Y' is open in Y, then the comorphism $\pi^*: k[Y'] \to k[\pi^{-1}(Y')]$ is an isomorphism of k[Y'] onto $k[\pi^{-1}(Y')]^G$; (iii) if W is a closed G-invariant subset of X, then $\pi(W)$ is closed in Y; (iv) if W_1 and W_2 are disjoint closed G-invariant subsets of X, then $\pi(W_1)$ and (W_2) are disjoint closed subsets of Y.*

In the case char k = 0, (a) is due mainly to H. Weyl and M. M. Schiffer, [116; Section 7, p.300] or [54; pp. 92-94]. In arbitrary characteristic, (a) is due primarily to C. Chevalley, D. Mumford, M. Nagata, and W. Haboush. An argument which only assumes facts about the Steinberg representations may be found in [100; Theorem 3.1.13, p.48]. All details may be found in [57]. The history of "Hilbert's fourteenth problem", is summarized in [77; pp. 90-92]. Statement (b) is a result from modern geometric invariant theory; a proof may be found in [77; Theorem 3.5, p.61] or [74; Theorem 1.1, p.27 and Theorem A.1.1, p.194].

In presenting the theory described above, I have tried to make the exposition throughout as complete as possible. This means that I have freely assumed any result stated or proved in [10] and, then, have tried to give almost all proofs. On occasion, proofs may be simplified if the assumption that k is of characteristic 0 is added. For then, finite-dimensional representations of reductive groups are known to be "completely reducible" [54; p.92]. In these cases, the simple proofs have been given assuming basic knowledge of this theory. However, they are not essential to the main flow of the exposition.

Notation.

We shall follow the notation introduced above throughout this book. In particular, G will always denote a linear algebraic group (with identity element e) and A a commutative k-algebra on which G acts rationally. If $a \in A$, the linear span of all the elements $g{\cdot}a$, $g \in G$, will be denoted by $\langle G{\cdot}a \rangle$. The symbol $\otimes$ will always mean $\otimes_k$ unless otherwise noted. The action of G on $k[G]$ by right (resp. left) translation will be denoted by r_g or $r(g)$ (resp. ℓ_g or $\ell(g)$). Thus, for $g,x \in G$ and $f \in k[G]$, we have $(r_g{\cdot}f)(x) = f(xg)$ (resp. $(\ell_g{\cdot}f)(x) = f(g^{-1}x)$).

A k-vector space V on which G acts as a group of automorphisms is called a *rational G-module* (or simply a "*G-module*") if it is a sum of finite-dimensional, G-invariant subspaces on each of which the given action of G is by some rational representation. If V is a G-module and $v \in V$, then the subgroup $G_v = \{g \in g : g{\cdot}v = v\}$ is the *stabilizer of* v and $G{\cdot}v = \{g{\cdot}v : g \in G\}$ is the *orbit of* v. If H is a subgroup of G, we put $V^H = \{v \in V : h{\cdot}v = v \text{ for all } h \in H\}$.

In general, if X is an algebraic variety over k, then we denote by $k[X]$ the k-algebra consisting of regular functions on X. Let G act morphically on X via a mapping $G \times X \rightarrow X$ denoted by $(g,x) \rightarrow g{\cdot}x$. Then G acts on $k[X]$ by $(g{\cdot}f)(x) = f(g^{-1}{\cdot}x)$ for all $g \in G$, $x \in X$, and $f \in k[X]$. In this way, the algebra $k[X]$ becomes a rational G-module. (The case where X is affine is proved in [10; Proposition 1.9, p.54]. In general, one needs that $k[G \times X] = k[G] \otimes k[X]$ and this result holds for arbitrary varieties X by [97; Theorem 1, p.400].) If Z is a subset of X, we shall denote the Zariski closure of Z by cl(Z). Other notation in the case where G is reductive is given at the beginning of §3. Whenever possible, we have tried to follow the notations and conventions of [10] except for denoting the underlying algebraically closed field by k which is now common in the literature.

Acknowledgements.

Many thanks to the referee whose suggestions greatly improved these notes. Also to Frédéric Bien and Armand Borel who invited me to their seminar in 1992 held at the Institute for Advanced Study on topics related to §23. I am grateful to Ferna Majewicz who typed an early draft of the manuscript. Finally, I dedicate this work to my wife Maxine, for her constant help with the manuscript and much more, and to my daughter Laura, for much sunshine.

Chapter One

Observable Subgroups

Introduction. Let H be a closed algebraic subgroup of G. In this chapter, we study the algebra $k[G/H]$ and give a geometric condition for it to be finitely generated over k. In §1, we define the notion of an "observable subgroup" of G and show that in studying $k[G/H]$, we may always assume that H is observable. Observable subgroups are defined by a number of equivalent conditions, some of interest in representation theory (§2). For example, if H is observable in G, then any finite-dimensional representation of H extends to a finite-dimensional representation of G. In addition, we shall see that H is observable if and only if the homogeneous space G/H may be embedded as an open set in an affine variety on which G acts morphically. Then, the algebra $k[G/H]$ is finitely generated if and only if such an embedding may be found so that the complement of the open subset G/H has codimension 2 (§4). In §5, we show that a maximal unipotent subgroup satisfies this property. Observable subgroups of a reductive group G which are normalized by a maximal torus in G are studied in §3.

§1. Stabilizer Subgroups

We begin this section by defining a correspondence between subgroups H of G and subalgebras R of $k[G]$. For H a subgroup of G (denoted by $H < G$), we define $H' = k[G]^H = \{f \in k[G] : r_h f = f \text{ for all } h \in H\}$. Then H' is a subalgebra of $k[G]$. For R a subalgebra of $k[G]$ (denoted by $R < k[G]$), we put $R' = \{g \in G : r_g f = f \text{ for all } f \in R\}$. Then R' is an algebraic subgroup of G. The basic properties of this "priming" correspondence are as follows.

Lemma 1.1. (i) *If $H_1 < H_2 < G$, then $H_2' < H_1'$. If $R_1 < R_2 < k[G]$, then $R_2' < R_1'$.*
(ii) *If $H < G$, then $H < H''$. If $R < k[G]$, then $R < R''$.*
(iii) *For all $H < G$, we have $H' = H'''$. For all $R < k[G]$, we have $R' = R'''$.*
(iv) *$G = G''$ and $\{e\} = \{e\}''$.*
(v) *If H and K are subgroups of G such that K normalizes H, then H' is invariant under K.*
(vi) *If H_1 and H_2 are closed subgroups of G with $H_1 < H_2 < G$ and H_2/H_1 an irreducible complete variety, then $H_1' = H_2'$.*
(vii) *Let H be a subgroup of G. There exist elements $f_1,\ldots,f_n \in H'$ such that $H'' = k[f_1,\ldots,f_n]'$, i.e., $H'' = \{g \in G : r_g f_i = f_i \text{ for } i = 1,\ldots,n\}$.*

Proof. Properties (i) and (ii) follow immediately from the definitions. Then since $H < H''$, it follows that $H''' < H'$. Furthermore, $H' < (H')'' = H'''$ and, so, $H' = H'''$. Similarly, $R' = R'''$. This proves statement (iii). Since $G \subset G''$, it follows that $G = G''$. If an element g is in $\{e\}''$, then r_g fixes each $f \in \{e\}' = k[G]$. Thus $f(g) = f(e)$ for all $f \in k[G]$ and $g = e$. This proves (iv). Next, let $h \in H$, $g \in K$, and $f \in H'$.

Let $m = g^{-1}hg$. Then, $r_h r_g f = r_g r_m f = r_g f$ since $m \in H$. To prove statement (vi), it suffices (by (i)) to show that $H_1' < H_2'$. Let $f \in H_1'$. Then the mapping $g \to g{\cdot}f$ gives rise to a morphism from H_2/H_1 into some finite-dimensional subspace of $k[G]$ (since G acts rationally on $k[G]$). Since H_2/H_1 is complete and irreducible, this morphism is constant, i.e., $f \in H_2'$. Finally, we prove statement (vii). For $f \in k[G]$, let us put $S(f) = k[f]' = \{g \in G : r_g f = f\}$. Now, $H'' = \cap\, S(f)$, where the intersection is taken over all $f \in H'$. We need to find $f_1,\ldots,f_n \in H'$ so that $H'' = \cap S(f_i)$. Suppose that this is false; we shall construct an infinite descending chain of algebraic groups $G \supsetneq H_1 \supsetneq H_2 \supsetneq \ldots \supsetneq H_n \supsetneq \ldots$ such that each H_j is a finite intersection of $S(f)$'s, $f \in H'$, and each H_j contains H''. Indeed, let $H_1 = S(f_1)$ where f_1 is any element in H'. We assume that $H_1,\ldots,H_n$ have been constructed and construct H_{n+1}. There is an $f \in H'$ so that $H_n \cap S(f) \subsetneq H_n$ for otherwise $H_n = H''$. We put $H_{n+1} = H_n \cap S(f)$. QED

Definition. Let H be a subgroup of G. The group H'' is called the *observable hull* (or *observable envelope*) of H in G. If $H = H''$, then H is said to be *observable* in G. If H is a closed subgroup of G and if $H'' = G$, then H is said to be *epimorphic* in G.

According to property (vi), any parabolic subgroup of (a connected group) G is epimorphic in G. Our attention now will be directed towards observable subgroups of G. We note that according to properties (i) to (iii) above, H'' is the smallest observable subgroup of G which contains H. That is, if L is observable in G and contains H, then L contains H'' (Exercise 1, below).

Theorem 1.2. *For H a closed subgroup of G, the following three conditions are equivalent:*
(i) $H = H''$;
(ii) *there is a finite-dimensional rational representation $\rho: G \to GL(V)$ and a vector $v \in V$ such that $H = G_v = \{g \in G : \rho(g)v = v\}$;*
(iii) *there are finitely many functions $f \in k[G/H]$ which separate the points in G/H.*
Proof. To show that condition (i) implies (ii), we apply statement (vii) above. Let $f_1,\ldots f_n \in H'$ be such that $H = H'' = k[f_1,\ldots f_n]'$. For each $i = 1,\ldots,n$, let V_i be a finite-dimensional subspace of $k[G]$, invariant under right translation by G, which contains f_i. Let $V = V_1 \oplus \ldots \oplus V_n$ and let $v = f_1 \oplus \ldots \oplus f_n$. Then $H = G_v$ and the cosets of G/H are in one-to-one correspondence with points in the orbit of v via the mapping $gH \to g{\cdot}v$. To show that condition (ii) implies (iii), let φ: $k[V] \to k[G]$ be defined by $(\varphi f)(g) = f(g{\cdot}v)$. Since $H = G_v$, we see that $\varphi f \in k[G/H]$. If $k[V] = k[x_1, \ldots, x_r]$, then the x_i separate points in V and the φx_i separate the cosets gH in G/H. Finally, suppose that there are functions $f_1, \ldots, f_r \in k[G/H] = k[G]^H$ which separate the cosets in G/H. Let $g \in H''$; then $r_g f_i(e) = f_i(e)$ for all i implies that $f_i(gH) = f_i(H)$ and, so, $g \in H$. QED

Corollary 1.3. *Let H be a closed subgroup of G. Then H is observable in G if and only if $H \cap G^o$ is observable in G^o.*
Proof. Let H be observable in G. Then, by the theorem, there is a finite-dimensional rational representation ρ: $G \to GL(V)$ and a vector $v \in V$ such that $H = G_v = \{g \in G : \rho(g)v = v\}$. It follows that $H \cap G^o$ is the stabilizer in G^o of v and, so, by the theorem again, $H \cap G^o$ is observable in G^o. Conversely, suppose that $H \cap G^o$ is observable in

G^o. Consider the coset decomposition of G with respect to G^o, distinguishing between the cosets which contain an element in H and those which do not: $G = \cup_i G^o h_i \cup \cup_j G^o g_j$, where $h_i \in H$ and $G^o g_j \cap H = \varnothing$ for all j. Let $f \in k[G^o/(H \cap G^o)]$ and define $F \in k[G]$ by $F(gh_i) = f(g)$ for $g \in G^o$ and $F = 0$ on each coset $G^o g_j$. We note that if $g,g' \in G^o$ and $h,h' \in H$ satisfy $gh = g'h'$, then $F(gh) = F(g'h') = f(g) = f(g')$ since f is invariant under $H \cap G^o$. It follows that $F \in k[G/H]$ and that F restricts to f. Choose non-zero functions $f_1, \ldots, f_m \in k[G^o/(H \cap G^o)]$ such that $(H \cap G^o) = \{g \in G^o : r_g f_i = f_i$ for all $i = 1, \ldots, m\}$; for each f_i, define $F_i \in k[G]$ as above. We shall show that $H = k[F_1, \ldots, F_m]'$; then H will be observable by property (iii), Lemma 1.1. As just shown, each $F_i \in k[G/H]$ so it is enough to prove that if $x \in G$ satisfies $r_x F_i = F_i$ for all i, then $x \in H$. We distinguish two cases. First, suppose that $x = g^o g_j \in G^o g_j$. There is an element $g' \in G^o$ with $f_1(g') \neq 0$. Then $0 \neq f_1(g') = F_1(g') = (r_x F_1)(g') = F_1(g' g^o g_j) = 0$, a contradiction. Second, suppose that $x = g^o h_i$ where g^o is not in $H \cap G^o$. Then g^o does not fix f_1, say. Let $y \in G^o$ be chosen so that $f_1(y) \neq f_1(yg^o)$. Then $f_1(y) = F_1(y) = (r_x F_1)(y) = F_1(yg^o h_i) = f_1(yg^o)$, another contradiction. QED

Using condition (ii) in the theorem and a basic result from algebraic groups, we are able to give many examples of observable groups.

Lemma 1.4 [10; 5.3]. *Let G be a linear algebraic group and let H be a closed subgroup of G.*
(i) *There is a character χ of H and functions $f_1,\ldots,f_n$ in $k[G]$ so that $H = \{g \in G : r_g f_i \in kf_i$ for each $i = 1,\ldots,n\}$ and $r_h f_i = \chi(h) f_i$ for all $h \in H$ and each $i = 1,\ldots,n$.*
(ii) *There is a finite-dimensional rational representation $\rho: G \rightarrow GL(V)$ and a vector $v \in V$ so that $H = \{g \in G : \rho(g)v \in kv\}$ and $\rho(h)v = \chi(h)v$ for all $h \in H$, where $\chi \in X(H)$.*

Corollary 1.5. *The subgroup $H_1 = \{h \in H : \chi(h) = 1\}$ is observable in G. In particular, if H has no non-trivial character, then H is observable in G.*
Proof. Obviously, $H_1 = G_v$ and we may apply the theorem. QED

It follows from the Corollary that the set of observable subgroups is fairly rich in G. For example, any connected unipotent or semi-simple subgroup of G is observable.

Corollary 1.6. *Let A be any commutative k-algebra on which G acts rationally. Then $A^H = A^{H''}$.*
Proof. Since $H \subset H''$, we have $A^{H''} \subset A^H$. Now let $f \in A^H$. According to the theorem, the subgroup $L = \{g \in G : g \cdot f = f\}$ is observable in G. Since L contains H, it also contains H''. QED

The Corollary shows the importance of observable subgroups in invariant theory. For if G acts rationally on a commutative k-algebra A and if H is a subgroup of G, then in studying A^H we may assume that H is observable. For the case where H is an epimorphic subgroup of a reductive group G, we obtain the following result by applying Theorem A.

Corollary 1.7. *Let G be reductive and act rationally on a commutative k-algebra A which is finitely generated over k. Let H be an epimorphic subgroup of G. Then $A^H = A^G$ is finitely generated over k.*

We conclude this section by making precise our geometric interpretation of observability. We begin with two elementary but very useful results.

Lemma 1.8. *Let H be a closed subgroup of G^o. Then $k[G^o]^H$ is integrally closed.*
Proof. Each point in G^o is simple (since G^o acts transitively on itself by right translation) so $k[G^o]$ is integrally closed. Let Q be the quotient field of $k[G^o]^H$. If $f \in Q$ is integral over $k[G^o]^H$, then $f \in k[G^o] \cap Q = k[G^o]^H$. QED

Lemma 1.9. *Let H be a closed subgroup of G having finite index. Then every finite-dimensional H-module is contained as an H-submodule in a finite-dimensional G-module.*
Proof. Let E be a finite-dimensional H-module. Let $A(H)$ (resp. $A(G)$) be the group algebra of H (resp. G) over k. Then $A(G)$ is a unitary right $A(H)$-module. Let $E_1 = A(G) \otimes_{A(H)} E$; then $A(G)$ acts on itself by left translation and this gives an action of $A(G)$ and G on E_1. The vector space E_1 is a finite-dimensional G-module which contains E (i.e, $A(H) \otimes_{A(H)} E$) as an H-submodule. QED

Before stating the next theorem, we need to recall the concept of normalization and two other basic facts from [10]. Let X be an irreducible variety with function field $k(X)$ and let L be a finite field extension of $k(X)$. There is a normal variety Y and a morphism π: $Y \to X$ such that (i) $k(Y) = L$, (ii) π is surjective and finite, (iii) if X' is any open affine subset of X, then $\pi^{-1}(X')$ is affine and $k[\pi^{-1}(X')]$ is the integral closure of $k[X']$ in L. The variety Y is called the *normalization* of X in L; it is known that if X is affine (resp. projective), then Y is also affine (resp. projective) [10; AG18.2]).

Theorem 1.10 [10; Theorem, p.43]. *Let α: $V \to W$ be a dominant morphism of irreducible normal varieties. Assume that the fibers of α have finite constant cardinality n. Then V is the normalization of W in $k(V)$ and n is the separable degree of $k(V)$ over $k(W)$. In particular, if α is birational then α is an isomorphism and if α is bijective then $k(V)$ is purely inseparable over $k(W)$.*

Theorem 1.11 [10; Proposition 6.7, p. 98]. *Let G act morphically on a variety V and let $x \in V$. Then $G{\cdot}x$ is a smooth variety, is locally closed in V, and the mapping π: $G \to G{\cdot}x$, $g \to g{\cdot}x$, is an orbit map for the action of G_x on G by right translation. Furthermore, π is a quotient of G by G_x if and only if π is separable, i.e., $(d\pi)_e$: $L(G) \to T(G(x))_x$ is surjective.*

Theorem 1.12. *Let H be an observable subgroup of G. There is a finite-dimensional rational representation ρ: $G \to GL(V)$ and a vector $v \in V$ such that $H = G_v$ and G/H is isomorphic to the orbit $G{\cdot}v = \{\rho(g)v : g \in G\}$.*
Proof. First, let us assume that G is connected. Let V, v, and ρ be as in Theorem 1.2. Let Z be the closure of the orbit $G{\cdot}v$. The mapping α: $G \to Z$, $g \to g{\cdot}v = \rho(g)v$, gives a G-equivariant embedding of $k[Z]$ in $k[G]$ where G operates on $k[G]$ by left translation

(Exercise 3). We shall consider $k[Z]$ to be a subalgebra of $k[G]$.

The morphism α gives rise to a morphism φ: $G/H \to Z$, $gH \to g{\cdot}v$. Since φ is injective and dominant, $k(G/H) = k(G)^H$ is a finite (algebraic) extension of $k(Z)$. Let X be the normalization of Z in $k(G/H)$; thus, $k[X]$ is the integral closure of $k[Z]$ in $k(G/H)$. Since $k[Z]$ is invariant under left translation, so is (its integral closure) $k[X]$. Furthermore, $k[X]$ is contained in $k[G] \cap k(G)^H = k[G]^H$. The G-equivariant inclusions $k[Z] \subset k[X] \subset k[G]^H \subset k[G]$ give rise to G-equivariant morphisms π: $X \to Z$ and ψ: $G/H \to X$ such that $\pi \circ \psi = \varphi$. The morphism ψ is injective since $\pi \circ \psi = \varphi$ is injective. Furthermore, if $\psi(eH) = x$, then $\psi(gH) = g{\cdot}x$ since ψ is G-equivariant. We may now apply Theorem 1.10 to see that G/H is isomorphic to the orbit $G{\cdot}x$.

Now, suppose that G is not connected. Consider the coset decomposition of G with respect to G^o, distinguishing between the cosets which contain an element in H and those which do not: $G = \cup_i G^o h_i \cup \cup_j G^o g_j$, where $h_i \in H$, $h_1 = e$ and $G^o g_j \cap H = \varnothing$ for all j. The group $H \cap G^o$ is observable in G^o (by Corollary 1.3). By what was just proved, there is a finite-dimensional rational representation ρ: $G^o \to GL(V)$ and a vector $v \in V$ such that $H \cap G^o = (G^o)_v$ and $G^o/(H \cap G^o)$ is isomorphic to the orbit $G^o{\cdot}v = \{\rho(g)v : g \in G^o\}$. We extend this representation of G^o to a representation of G on a vector space V_1 as in Lemma 1.9; then $V_1 = A(G) \otimes_B V$ where $B = A(G^o)$. The algebra $A(G)$ is a free B-module with basis consisting of all the h_i and g_j. By a fundamental property of tensor products [55; Theorem 5.11, p.215], each $v_1 \in V_1$ can be written uniquely in the form $\Sigma_i h_i \otimes_B v_i + \Sigma_j g_j \otimes_B v_j$ and, so, V_1 is a direct sum of vector spaces each isomorphic to V. If $g \in G^o$, we have $g{\cdot}(\Sigma_i h_i \otimes_B v_i + \Sigma_j g_j \otimes_B v_j) = \Sigma_i gh_i \otimes_B v_i + \Sigma_j gg_j \otimes_B v_j = \Sigma_i h_i(h_i^{-1}gh_i) \otimes_B v_i + \Sigma_j g_j(g_j^{-1}gg_j) \otimes_B v_j = \Sigma_i h_i \otimes_B (h_i^{-1}gh_i)v_i + \Sigma_j g_j \otimes_B (g_j^{-1}gg_j)v_j$.

Let $z = \Sigma_i h_i \otimes_B v$. We will show that G/H is isomorphic to the orbit $G{\cdot}z$. First, $G^o{\cdot}z$ is isomorphic to $G^o{\cdot}v$. Indeed, the mapping from G^o to $G^o{\cdot}z$ given by $g \to g{\cdot}z = \Sigma_i h_i \otimes_B (h_i^{-1}gh_i)v$ factors through $G^o/(H \cap G^o)$. Since $G^o{\cdot}v$ is isomorphic to $G^o/(H \cap G^o)$, it also factors through $G^o{\cdot}v$. Its inverse is the projection mapping $G^o{\cdot}z \to G^o{\cdot}v$, $g{\cdot}z = \Sigma_i h_i \otimes_B (h_i^{-1}gh_i)v \to h_1 \otimes_B g{\cdot}v = e \otimes_B g{\cdot}v \to g{\cdot}v$. Secondly, $H \subset G_z$. For if $h \in H$, then $h{\cdot}z = \Sigma_i hh_i \otimes_B v = \Sigma_i h_i h_i' \otimes_B v$ for suitable elements $h_i' \in H \cap G^o$. But this last expression is $\Sigma_i h_i \otimes_B h_i'v = \Sigma_i h_i \otimes_B v = z$ by our choice of v. Furthermore, $G_z \subset H$. Indeed, since the representation of G extends that of G^o, we see that $G_z \cap G^o \subset (G^o)_z = H \cap G^o$. Now suppose that g fixes z but is not in H. Then g may be written in the form $g^o g_j$ for a suitable $g^o \in G^o$ and some j. Now, $z = g{\cdot}z$ implies that $\Sigma_i h_i \otimes_B v = \Sigma_i gh_i \otimes_B v \in \Sigma_j g_j \otimes_B V$. The unique representation of each $v_1 \in V_1$ noted above shows that the last equality cannot hold. To complete the proof, we need to show that G/H is isomorphic to $G{\cdot}z$. The mapping $G^o \to G^o{\cdot}z$, $g \to g{\cdot}z$, is separable by assumption; thus, the mapping $G \to G{\cdot}z$ is also separable. This shows that G/H is isomorphic to $G{\cdot}z$ (by Theorem 1.11). QED

Exercises.

1. Let H be a subgroup of G. Show that H'' is the smallest observable subgroup which contains H.

2. Let H_1 and H_2 be observable subgroups of G. Show that $H_1 \cap H_2$ is an observable subgroup of G.

3. Let G act morphically on a variety X. Let $x \in X$ and let φ: $k[X] \to k[G]$ be given

by $(\varphi f)(g) = f(g{\cdot}x)$. Let $G_x = \{g \in G : g{\cdot}x = x\}$ and let R be the image of $k[X]$ under φ. Show that the mapping φ is G-equivariant from $k[X]$ to $k[G/G_x]$, where G acts on $k[G]$ by left translation.

4. Let $G = SL_2(k)$, i.e., the set of all 2 x 2 matrices (x_{ij}) with entries in k and having determinant 1. Let $U = \{(x_{ij}) \in G : x_{11} = x_{22} = 1 \text{ and } x_{21} = 0\}$. Show that U is observable in G by finding a point v in the plane so that $U = G_v$.

5. Let H be a closed subgroup of G. If H is an epimorphic subgroup of G, show that $H \cap G^o$ is an epimorphic subgroup of G^o. Conversely, if $H \cap G^o$ is an epimorphic subgroup of G^o and if each coset of G/G^o contains an element in H, then H is epimorphic in G.

6. Let H be a normal subgroup of G. Let $f \in k[G]$. Show that f is fixed under left translation by H if and only if it is fixed under right translation by H.

§2. Equivalent Conditions

For $H < G$ to be observable, we initially required that $H = H''$, a condition involving invariant functions. This fact turned out to be equivalent to H being a stabilizer subgroup (Theorem 1.2). Originally, however, the concept of an "observable subgroup" was introduced by A. Bialynicki-Birula, G. Hochschild, and G.D. Mostow because of its interest in representation theory. We next give this interpretation along with several others of importance.

Theorem 2.1. *For H a closed subgroup of G, the following conditions are equivalent.*
(1) $H = H''$.
(2) There is a finite-dimensional rational representation $\rho: G \to GL(V)$ and a vector $v \in V$ so that $H = G_v = \{g \in G : \rho(g)v = v\}$.
(3) There are finitely many functions $f \in k[G/H]$ which separate the points in G/H.
(4) G/H is quasi-affine, i.e., is isomorphic to an open subset of an affine variety.
(5) The quotient field of $k[G^o/(H \cap G^o)]$ is the subfield of $k(G^o)$ fixed under right translation by $H \cap G^o$.
(6) For every 1-dimensional rational H-module that is contained as an H-submodule of some finite-dimensional rational G-module, the dual H-module is also an H-submodule of a finite-dimensional G-module.
(7) Every finite-dimensional rational H-module is an H-submodule of a finite-dimensional rational G-module.

Proof. The equivalence of statements (1), (2), and (3) follows from Theorems 1.2. Also, statement (2) implies (4) by Theorem 1.12 and (4) obviously implies (3). (4) ⇒ (5). The subgroup $H \cap G^o$ is observable in G^o by Corollary 1.3. We now apply Theorem 1.12. Let X be an affine variety and suppose that there is an open dense subset X' of X such that $G^o/(H \cap G^o)$ is isomorphic to X'. If Q is the quotient field of $k[G^o/(H \cap G^o)]$, then the subfield of $k(G^o)$ fixed under right translation by $H \cap G^o$ is $k(G^o/(H \cap G^o)) = k(X') = k(X) =$ the quotient field of $k[X] \subset$ the quotient field of $k[X'] = Q$.

(5) ⇒ (6). First, let us note that the homogeneous spaces $G^o/(H \cap G^o)$ and G^oH/H are isomorphic. (The natural mapping $G^o \to G^oH \to G^oH/H$ factors through $G^o/(H \cap G^o)$.)

Let V be a finite-dimensional, rational G-module and let $v \in V$ be a non-zero vector such that $h{\cdot}v = \chi(h)v$ for all $h \in H$, where χ is a character of H. We now prove (6) when G is replaced by G^oH; then, we apply Lemma 1.9 to obtain the statement for G. Let λ be any element in the dual space of V so that $\lambda(v) \neq 0$ and let $f \in k[G^oH]$ be defined by $f(g) = \lambda(g{\cdot}v)$. Then $f(e) \neq 0$ and $r_h{\cdot}f = \chi(h)f$ for all $h \in H$. For every $x \in G^oH$, we have ℓ_xf/f is in $k(G^oH/H)$ since $r_h(\ell_xf/f)(g) = (\ell_xf/f)(gh) = f(x^{-1}gh)/f(gh) = \chi(h)f(x^{-1}g)/\chi(h)f(g) = (\ell_xf)(g)/f(g)$. According to statement (5), there are non-zero elements a_x, $b_x \in k[G^oH/H]$ such that $(\ell_xf)b_x = a_xf$. Let Z be the set of all zeros of f on G^oH and let Z' be its complement. We note that Z' is an open, dense subset of G^oH since $r_h{\cdot}f = \chi(h)f$ for all $h \in H$. Let $g \in Z \cap xZ'$. Then $(\ell_xf)(g)b_x(g) = a_x(g)f(g) = 0$ since $g \in Z$. However, $(\ell_xf)(g) = f(x^{-1}g) \neq 0$ since $g \in xZ'$. We may conclude that $b_x(g) = 0$, i.e., b_x vanishes on $Z \cap xZ'$. The open sets yZ', $y \in G^oH$, cover G^oH. Hence, there are elements $x(1),\ldots,x(n)$ in G^oH so that Z is the union of the sets $Z \cap x(i){\cdot}Z'$. Let $b = b_{x(1)}\ldots b_{x(n)}$. Then b is a non-zero element in $k[G^oH/H]$ which vanishes on Z. According to Hilbert's Nullstellensatz, there is a positive integer m so that $b^m = cf$ where $c \in k[G^oH]$. Then, for each $h \in H$, $cf = b^m = (r_h b)^m = (r_h c)(r_h f) = (r_h c)(\chi(h)f)$ from which we see that $r_h c = \chi(h)^{-1}c$ since Z' is open, dense. This proves statement (6).

(6) ⇒ (7). First, let us prove that any finite-dimensional rational H-module may be identified with an H-submodule of a direct sum of a finite number of copies of $k[H]$, on which H acts by right translation. Indeed, let V be a finite-dimensional rational H-module having basis $\{v_1,\ldots v_n\}$ and let us suppose that f_{ij}, $1 \leq i,j \leq n$, are elements in $k[H]$ so that $h{\cdot}v_j = \Sigma_{i=1}^n f_{ij}(h)v_i$ for all $h \in H$. We identify v_j with $(f_{1j},\ldots f_{nj})$ and check (Exercise 5) that this mapping is H-equivariant.

The restriction mapping $k[G] \to k[H]$ gives rise to a surjective, H-equivariant (where H acts by right translation) mapping π from $k[G] \oplus\ldots\oplus k[G]$ (n copies) to $k[H] \oplus\ldots\oplus k[H]$ (n copies). Thus, V may be identified with a quotient W/Z where W is a finite-dimensional, rational H-module that is contained as an H-submodule in a finite-dimensional, rational G-module M, and where Z is an H-submodule of W.

Let $\Lambda(M)$ be the exterior algebra over M. Let $m = \dim Z$; obviously $\Lambda^{m+1}(M)$ contains $W \wedge \Lambda^m(Z)$ as an H-submodule. Let $\{z_1,\ldots,z_m,w_1,\ldots,w_n\}$ be a basis for W such that $\{z_1,\ldots,z_m\}$ is a basis of Z and $\pi(w_i) = v_i$. The H-modules $W \wedge \Lambda^m(Z)$ and $V \otimes \Lambda^m(Z)$ are isomorphic via the mapping $w_i \wedge (z_1 \wedge\ldots\wedge z_m) \to v_i \otimes (z_1 \wedge\ldots\wedge z_m)$. Since Z is H-invariant and $\dim Z = m$, there is a character μ of H such that $h(z_1 \wedge\ldots\wedge z_m) = \mu(h)z_1 \wedge\ldots\wedge z_m$ for all $h \in H$. According to (6), there is a finite-dimensional, rational G-module M_1 and a non-zero element $m_1 \in M_1$ such that $h{\cdot}m_1 = \mu(h)^{-1}m_1$ for all $h \in H$. Let $<m_1> = \{cm_1 : c \in k\}$. Then, V is isomorphic as an H-module to $V \otimes \Lambda^m(Z) \otimes <m_1>$ which is isomorphic to $(W \wedge \Lambda^m(Z)) \otimes <m_1>$, an H-submodule of the G-module $\Lambda^{m+1}(M) \otimes M_1$.

(7) ⇒ (1). According to Lemma 1.4, there is a finite-dimensional rational representation $\rho: G \to GL(V)$, a vector $v \in V$ and a character χ of H so that $H = \{g \in G : \rho(g)v \in kv\}$ and $\rho(h)v = \chi(h)v$ for all $h \in H$. Applying (7), we may find a finite-dimensional, rational G-module W and a non-zero element $w \in W$ so that $h{\cdot}w = \chi(h)^{-1}w$ for all $h \in H$. Then, H is the stabilizer in G of the vector $v \otimes w$ in $V \otimes W$. This proves (2) and, hence, (1). QED

Corollary 2.2. *Let H be a closed subgroup of G. The following conditions are equivalent:* (i) *H is observable in G,* (ii) *$H \cap G^o$ is observable in G^o,* (iii) *H^o is observable in G^o.*
Proof. We have shown that (i) and (ii) are equivalent (Corollary 1.3). Hence, in proving that (i) and (iii) are equivalent, we may assume that G is connected. First, suppose that H is observable in G. The mapping $G/H^o \to G/H$, $gH^o \to gH$, has finite fibers, all of cardinality H/H^o. Thus, using Theorem 1.10, we see that G/H^o is the normalization of G/H in $k(G/H^o)$. On the other hand, according to Theorem 1.12, there is a finite-dimensional rational representation ρ: $G \to GL(V)$ and a vector $v \in V$ such that $H = G_v$ and G/H is isomorphic to $G{\cdot}v$. Let $X = \mathrm{cl}(G{\cdot}v)$ and let Y be the normalization of X in $k(G/H^o)$. Let π: $Y \to X$ be the natural mapping. By the definition of normalization, $\pi^{-1}(G{\cdot}v)$ is the normalization of $G/H = G{\cdot}v$ in $k(G/H^o)$. Since $\pi^{-1}(G{\cdot}v)$ is quasi-affine and is isomorphic to G/H^o, we may apply (4), Theorem 2.1.

Next, suppose that H^o is observable. Let f be an element in $k(G/H)$. According to (5), Theorem 2.1, we need to show that $f = a/b$ where $a,b \in k[G]^H$. Since H^o is observable, we have $f = a_1/b_1$ where a_1, $b_1 \in k[G/H^o]$. Let $H = eH^o \cup x(1)H^o \cup \ldots \cup x(m)H^o$ be the distinct cosets of H^o in H and let $B = (r_{x(1)}b_1)\ldots(r_{x(m)}b_1)$. Then $f = a_1B/b_1B$. The element b_1B is in $k[G]^H$; since $f \in k(G)^H$, we see that a_1B is also in $k[G]^H$. QED

Corollary 2.3. *Let $H < K < G$. If H is observable in K and K is observable in G, then H is observable in G.*
Proof. This follows immediately from (7), Theorem 2.1. QED

Corollary 2.4. *If H is reductive, then H is observable in G.*
Proof. The subgroup H acts on G by right translation and the orbits of this action are the cosets of H in G. All of the orbits are closed since they have the same dimension. According to Theorem A, the algebra $k[G]^H$ is finitely generated over k. Let Y be the corresponding affine variety and let π: $G \to Y$ be the mapping corresponding to the inclusion of $k[G]^H$ in $k[G]$. Applying Theorem A, again, we see that the orbits of H on G go to distinct points of Y. Thus, H is observable in G by Theorem 2.1(3). (According to Theorem 2.1(4), G/H is quasi-affine. We shall see in Corollary 4.6 that G/H is actually the affine variety Y.) QED

Corollary 2.5. *Let G be solvable. Then any closed subgroup H of G is observable. In fact, the quotient space G/H is affine.*
Proof. To prove that H is observable, we may assume that both G and H are connected (Corollary 2.2). Let $G = TU$ (resp. $H = SV$) where T (resp. S) is a maximal torus in G (resp. H), $U = G_u$, and $V = H_u$. Now, $X(H) = X(S) \subset X(T) = X(G)$ so H is observable in G by Theorem 2.1 (6). Next, let W be a finite-dimensional G-module and $w \in W$ be such that $H = G_w$ and $G/H \simeq G{\cdot}w$ (Theorem 1.12). Let $X = \mathrm{cl}(G{\cdot}w)$ and $Y = X - G{\cdot}w$. Since G is solvable and Y is G-invariant, there is a non-zero $f \in k[X]$, which is 0 on Y, and a $\chi \in X(G)$ so that $g{\cdot}f = \chi(g)f$ for all $g \in G$. Then, $f(g{\cdot}w) = \chi(g^{-1})f(w) \neq 0$ and $Y = \{x \in X : f(x) = 0\}$. Then, $k[G/H] = k[G{\cdot}w] = k[X][1/f]$. QED

Lemma 2.6 ("rigidity of Tori", [10; Proposition, p.117]). *Let α: $V \times H \to L$ be a morphism of varieties such that:*
(i) *L is an algebraic group containing, for each $m > 0$, only finitely many elements of*

order m;
(ii) *H is an algebraic group in which the elements of finite order are dense; and*
(iii) *V is a connected variety and, for each* $x \in V$, α_x: $h \to \alpha(x,h)$ *is a homomorphism.*
Then the map $x \to \alpha_x$ *is constant.*

Theorem 2.7. *Let H be a closed subgroup of G. Then H is observable in G if and only if the radical of H is observable in G.*
Proof. We may assume that both H and G are connected (Corollary 2.2). Let R be the radical of H. Since R is normal in H, the quotient space H/R is affine. Thus, R is observable in H by Theorem 2.1, (4). If, in addition, H is observable in G, then R is observable in G by Corollary 2.3. Now, suppose that R is observable in G; let $R = TU$, where T is a maximal torus in R and U consists of the unipotent elements in R. Let χ be any character of R. Consider the mapping $H \times T \to k^*$ given by $(h,t) \to \chi(h^{-1}th)$; we apply Lemma 2.6 to see that $\chi(h^{-1}th) = \chi(t)$ for all $t \in T$ and $h \in H$. Thus, $\chi(h^{-1}rh) = \chi(r)$ for all $r \in R$ and $h \in H$. Next, according to Lemma 1.4(b), there is a finite-dimensional rational representation ρ: $G \to GL(V)$ and a vector $v \in V$ so that $H = \{g \in G : \rho(g)v \in kv\}$ and $\rho(h)v = \chi(h)v$ for all $h \in H$, where $\chi \in X(H)$. If $\chi \mid R$ is trivial, then χ is trivial on H since H/R is connected and semi-simple. Then, $H = G_v$ is observable in G by Theorem 1.2. Otherwise, by Theorem 2.1, (6), there is finite-dimensional G-module W and a non-zero element $w_o \in W$ with $r{\cdot}w_o = \chi(r)^{-1}w_o$ for all $r \in R$. We put $W(-\chi) = \{w \in W : r{\cdot}w = \chi(r)^{-1}w$ for all $r \in R\}$. Then $W(-\chi)$ is H-invariant. Indeed, if $w \in W(-\chi)$, then $r{\cdot}(h{\cdot}w) = h{\cdot}(h^{-1}rh){\cdot}w = h{\cdot}\chi(h^{-1}rh)^{-1}w = h{\cdot}\chi(r)^{-1}w$ as was noted above. Let $\{w_1,\ldots,w_m\}$ be a basis of $W(-\chi)$, let $V_1 = V \otimes \ldots \otimes V$ (m-times) $\otimes \Lambda^m(W)$ together with the natural representation ρ_1 of G, and let $v_1 = v \otimes \ldots \otimes v \otimes (w_1 \wedge \ldots \wedge w_m)$. We have seen that H maps $W(-\chi)$ to $W(-\chi)$; thus, $h{\cdot}(w_1 \wedge \ldots \wedge w_m) \in k(w_1 \wedge \ldots \wedge w_m)$ for all $h \in H$. Using this fact, we now show that H is the stabilizer of v_1 in G. First, let $g \in G$ stabilize v_1. Then $\rho(g){\cdot}v \in kv$ and, so $g \in H$. Second, let $h \in H$; we need to show that $\rho_1(h){\cdot}v_1 = v_1$. Since H maps $W(-\chi)$ into itself, there is a character $\mu \in X(H)$ so that $\rho_1(h){\cdot}v_1 = \mu(h)v_1$. Since $X(U) = \{0\}$, μ is trivial on U. Also, for all $t \in T$ we have $\mu(t) = \chi(t)^m\chi(t)^{-m} = 1$. Therefore, μ gives rise to a character on H/R. But H/R is connected and semi-simple, so, $\mu = 1$. QED

Corollary 2.8. *If the radical of H is unipotent, then H is observable in G.*
Proof. This follows from Theorem 2.7 and Corollary 1.5. QED

Corollary 2.9. *If the radical of H is nilpotent, then H is observable in G.*
Proof. See Exercise 6, below.

In stating the next corollary, we adapt the notation that $\mathfrak{R}_uH$ denotes the unipotent radical of the subgroup H.

Corollary 2.10. *Let L be an observable subgroup of G and let T be a maximal torus in L. Let H be a closed subgroup of L which is normalized by T and satisfies* $\mathfrak{R}_uH \subset \mathfrak{R}_uL$. *Then H is observable in G. In particular,* $T\mathfrak{R}_uL$ *is observable in G.*
Proof. First, let us show that $T\mathfrak{R}_uL$ is observable in L. Indeed, $L/(T\mathfrak{R}_uL) = (L/\mathfrak{R}_uL)/((T\mathfrak{R}_uL)/\mathfrak{R}_uL)$ and the latter is a quasi-affine variety (by Theorem 2.1(4)) since

$T\mathfrak{R}_uL/\mathfrak{R}_uL$ is a torus and so is observable in $(L/\mathfrak{R}_uL)$ by Corollary 2.4. Thus, $T\mathfrak{R}_uL$ is observable in L by Theorem 2.1(4) and in G by Corollary 2.3. Next, we show that H is observable in G. To do this, we may replace H by its radical R and also suppose that $R \subset T\mathfrak{R}_uH$. By Corollary 2.5, R is observable in $T\mathfrak{R}_uH$ and $T\mathfrak{R}_uH$ is observable in $T\mathfrak{R}_uL$. Thus, R is observable in G by Corollary 2.3. QED

In §7, we shall see that if $\mathfrak{R}_uH \subset \mathfrak{R}_uL$, then L/H is actually an affine variety.

Corollary 2.11. *Let H be an observable subgroup of G. Then $H{\cdot}\mathfrak{R}_uG$ is also observable in G.*

Proof. We may assume that both G and H are connected (Corollary 2.2). Since $\mathfrak{R}_uG$ is normal in G, an element $f \in k[G]$ is fixed under right translation by $\mathfrak{R}_uG$ if and only if it is fixed under left translation (Exercise 6, §1). Now, let $f \in k(G)$ be fixed under right translation by the group $H{\cdot}\mathfrak{R}_uG$. We shall show that Theorem 2.1(5) holds. First, applying Theorem 2.1(5) to H, we see that there are elements $a, b \in k[G]^H$ such that $f = a/b$. Let $V = \{c \in k[G]^H : cf \in k[G]\}$. Since $b \in V$, we see that $V \neq \{0\}$. Furthermore, V is invariant under left translation by $\mathfrak{R}_uG$. Thus, there is a non-zero element $v \in V$, which is fixed under left translation (and, so, under right translation) by $\mathfrak{R}_uG$. Suppose that $vf = w$. Then $f = v/w$ as desired. QED

Exercises.

1. Let H be a closed subgroup of G. Show that H is observable in G if any one of the following conditions holds: (i) H is normal in G; (ii) H is a torus; (iii) H is of finite index in G.
2. Give an example of a group G and a closed subgroup H so that G/H is quasi-affine, but is not affine.
3. Let $B = TU$ be a Borel subgroup of a connected group G, where T is a maximal torus and U is a maximal unipotent subgroup. Show that both T and U are observable in G, but that $B = TU$ is not.

4 [4; p.143]. Let $G = GL_2(k)$, i.e., the set of all 2 x 2 matrices (x_{ij}) with entries in k and having non-zero determinant. Let $L = \{(x_{ij}) \in G : x_{11} = a, x_{22} = a^2$ for some $a \in k^*$ and $x_{21} = 0$. Show that L is observable in G.

5. Show that the mapping given in the proof of Theorem 2.1, (6) ⇒ (7), which identifies v_j with $(f_{1j},\ldots f_{nj})$ is H-equivariant.
6. Prove Corollary 2.9. (Hint: proceed as in the proof of Theorem 2.7. Use the fact that a connected nilpotent group H can be written as $T \times H_u$.)
7. Let H be a subgroup of G. Show that $(H^o)'' = (H'')^o$.
8. Let H be an observable subgroup of G. Let W be a finite-dimensional H-module. Show that there is a finite-dimensional G-module V which contains W as an H-submodule and is such that the natural mapping $k[V] \rightarrow k[W]$ is H-equivariant.

§3. Observable Subgroups of Reductive Groups

Our goal in this section is to describe the observable subgroups of a reductive group G which are normalized by a maximal torus T (Theorem 3.9). Many of the preliminary results are of interest in themselves and will also be used in later sections. We begin by recalling some facts about representations of reductive algebraic groups [10, §24B or 54, Section 31]; these will be useful below and, also, in §5. The notation introduced here will be used throughout this section unless otherwise noted.

Let G be a connected, reductive algebraic group. Let T be a maximal torus in G and let $X(T)$ be its character group. The Weyl group of T will be denoted by $W(T, G)$. For H a closed subgroup of G normalized by T, we denote by $\Phi(T, H)$ the set of roots of H relative to T. In particular, for $\alpha \in \Phi(T, G)$, let x_α: $\mathbf{G}_a \to U_\alpha$ be the canonical isomorphism; thus, $tx_\alpha(c)t^{-1} = x_\alpha(c\alpha(t))$. Let B be a Borel subgroup of G which contains T and has maximal unipotent subgroup U. Let $\Phi^+ = \Phi(T,B)$; the elements in Φ^+ are called positive roots. Let $B^- = TU^-$ be the Borel subgroup of G which is "opposite" to B; then $\Phi^- = \Phi(T,B^-) = -\Phi^+$. Let $w_o = s_oT$ be the element in the Weyl group of T such that $s_oUs_o^{-1} = U^-$. We denote by ι the involution of $X(T)$ given by $\iota\chi = -s_o\chi$. A partial order may be defined on $X(T)$ by $\chi' \geq \chi$ if $\chi' - \chi$ is a sum of positive roots. An element $\omega \in X(T)$ is called *dominant* if $(\omega, \alpha) \geq 0$ for all $\alpha \in \Phi^+$. Let $X^+(T)$ be the set of all dominant weights. Let $E = \mathbf{R} \otimes_{\mathbf{Z}} X(T)$. The following results may then be proved.

(a) Let V be a finite-dimensional, irreducible rational G-module.
(i) There is a unique B-stable, one-dimensional subspace of V spanned by a vector v having weight $\omega \in X^+(T)$. The vector v is called a *highest weight vector* in V (with respect to the Borel subgroup B) and ω is called a *highest weight*.
(ii) If χ is any other weight of T on V, then $\omega - \chi$ is a sum of positive roots.
(iii) Let V' be another finite-dimensional, irreducible rational G-module and suppose that V' has highest weight ω'. Then V' is isomorphic to V if and only if $\omega' = \omega$.
(iv) If $\omega \in X^+(T)$, there is a finite-dimensional, irreducible G-module V having highest weight ω; the representation of G on the dual space V^* is irreducible and has highest weight $\iota\omega = -s_o\omega$.
(v) Suppose that G is semi-simple. Let $\Delta = \{\alpha_1, \dots, \alpha_r\}$ be the basis for Φ^+ corresponding to B. Let $\omega_1, \dots, \omega_r$ be the elements in E so that $2(\omega_i,\alpha_j)/(\alpha_j, \alpha_j) = \delta_{ij}$ for all $i,j = 1, \dots, r$. (The ω_i are called *fundamental weights* of G.) Then any dominant weight ω can be written uniquely in the form $\Sigma_i c_i\omega_i$ where the c_i's are non-negative integers.
(b) Let V be a rational G-module. If $\chi \in X(T)$, let $V(\chi) = \{v \in V : t\cdot v = \chi(t)v$ for all $t \in T\}$.
(i) If χ is a non-zero weight of T on a rational G-module V, then $2(\chi, \alpha)/(\alpha, \alpha) \in \mathbf{Z}$.
(ii) Let $w = sT \in W(T, G)$. For $\chi \in X(T)$, we define $w\chi$ by $w\chi(t) = \chi(s^{-1}ts)$ for all $t \in T$. Then $w\cdot V(\chi) = V(w\chi)$.
(c) Let G be semi-simple. There is a connected semi-simple algebraic group G_1 having a Borel subgroup $B_1 = T_1U_1$ and a surjective homomorphism π: $G_1 \to G$ having finite kernel such that:
(i) $X(T_1)$ has $\{\omega_1, \dots, \omega_r\}$ as a $\mathbf{Z}$-basis;
(ii) $\pi(U_1) = U$ and $\pi(T_1) = T$.
The group G_1 is called the *simply connected group* of type $\Phi(T, G)$.

Lemma 3.1. *Let V be a rational G-module and let $v \in V(\chi)$.*
(a) If $\alpha \in \Phi(T,G)$, then there are vectors $v_i \in V(\chi + i\alpha)$ $(i = 1,2,\dots)$ so that $x_\alpha(c)\cdot v$

$= v + \Sigma c^i v_i$ *for all* $c \in k$.
(b) *If* $\alpha_1, \ldots, \alpha_m \in \Phi(T,G)$, *then for all* $c_1, \ldots, c_m \in k$,

$$x_{\alpha_1}(c_1) \ \ldots \ x_{\alpha_m}(c_m)\cdot v = v + \sum c_1^{e_1} \ldots c_m^{e_m} v_{(e)}$$

where $v_{(e)}$ *has T-weight* $\chi + e_1\alpha_1 + \ldots + e_m\alpha_m$ *and the* $e_1, \ldots, e_m$ *are non-negative integers.*
Proof. The mapping from $\mathbf{G}_a$ to V given by $c \to x_\alpha(c)\cdot v$ is polynomial. Thus, there are vectors $v_i \in V$ for $i = 0,\ldots,n$, so that $x_\alpha(c)\cdot v = \Sigma_i c^i v_i$ for all $c \in \mathbf{G}_a$. Putting $c = 0$, we see that $v_0 = v$. Let $t \in T$. Then, $tx_\alpha(c)t^{-1} = x_\alpha(\alpha(t)c)$ and $\Sigma c^i(t\cdot v_i) = tx_\alpha(c)\cdot v = x_\alpha(\alpha(t)c)t\cdot v = \Sigma(\alpha(t)c)^i\chi(t)v_i$. Thus, $t\cdot v_i = (\chi + i\alpha)(t)v_i$. This proves statement (a) and then, statement (b) follows by induction on m. QED

Theorem 3.2. *Let v be a non-zero element in* $V(\chi)$. *Let* $\alpha \in \Phi(T,G)$ *and suppose that* $U_\alpha \subset G_v$. *Then* (i) $(\chi,\alpha) \geq 0$ *and* (ii) $(\chi,\alpha) = 0$ *if and only if* $U_{-\alpha} \subset G_v$.
Proof. Let H be the group generated by T, U_α and $U_{-\alpha}$ [10; Theorem, p.176]. Let V_1 be the subspace of V spanned by all elements of the form $h\cdot v$, $h \in H$. Then V_1 is also spanned by all elements of the form $h\cdot cv$, where $h \in U_{-\alpha}$ and $c \in k$, since H is contained in $TU_\alpha \cup U_{-\alpha}TU_\alpha$, $U_\alpha \subset G_v$, and $v \in V(\chi)$. According to Lemma 3.1, the weights of T on V_1 all have the form $\chi - r\alpha$ where r is a non-negative integer. Furthermore, $V_1(\chi) = V_1 \cap V(\chi)$ is a one-dimensional space spanned by v. Let w denote the reflection in the Weyl group of T which is determined by α; we may assume that $w \in H$. Now, to prove statement (i), we first observe that $w\chi$, which is known to be $\chi - (2(\chi,\alpha)/(\alpha,\alpha))\alpha$ ([54; p.189]), is a weight of T on V_1 since $wV(\chi) = V(w\chi)$. Since all the weights of T on V_1 have the form $\chi - r\alpha$, where r is a non-negative integer, we see that $(\chi,\alpha) \geq 0$. If, in addition, $U_{-\alpha}$ is contained in G_v, then $(\chi, -\alpha) \geq 0$ by (i) and, so, $(\chi,\alpha) = 0$. On the other hand, if $(\chi,\alpha) = 0$, then $w\chi = \chi$ and $wV_1(\chi) = V_1(\chi)$. Since $U_{-\alpha} = wU_\alpha w^{-1}$, we see that $U_{-\alpha}$ sends $V_1(\chi)$ to itself. However, since $U_{-\alpha}$ has no non-trivial character, $U_{-\alpha} \subset G_v$. QED

Lemma 3.3 [10; Proposition, p.182]. *Let H be a unipotent subgroup of G which is normalized by T. Then H is connected and is directly spanned by* $\{U_\alpha: \alpha \in \Phi(T, H)$ *where the* U_α*'s may be taken in any order.*

Theorem 3.4 [10; Proposition 13.20, p.178]. *Let L be a closed connected subgroup of G which is normalized by T. Let* L_1 *be the group generated by all the* U_α, $\alpha \in \Phi(T, L)$. *Then* $L = L_1 \cdot (T \cap L)^o$. *Furthermore,* $U_\alpha \subset L$ *if and only if* $\alpha \in \Phi(T, L)$.

It follows from the previous result that the unipotent radical of L is generated by those U_α such that $\alpha \in \Phi(T, L)$ but $-\alpha \notin \Phi(T, L)$ (Exercise 1).

Corollary 3.5. *Let H be a closed subgroup of G which contains U and is normalized by T. Let* H_1 *be the subgroup of H which is generated by all the* U_α, $\alpha \in \Phi(T, H)$. *Then* $H = H_1\cdot(T \cap H)$.
Proof. The group TH is a parabolic subgroup of G since it contains the Borel subgroup TU. Thus, TH is connected. If $U_\alpha \subset TH$, then $U_\alpha \subset H$ as may be seen by considering

the homomorphism from TH to a torus, $TH \to TH/H$. According to Theorem 3.4, $TH = H_1 \cdot T$. Hence, $H = H_1 \cdot (T \cap H)$. QED

Corollary 3.6. *Let $\chi \in X(T)$. Let P_χ be the group generated by all the U_α such that $(\chi, \alpha) \geq 0$ and $\{t \in T : \chi(t) = 1\}$. Then $P = TP_\chi$ is a parabolic subgroup of G and P_χ is the stabilizer of a highest weight vector. Furthermore, the stabilizer of any highest weight vector with respect to B has the form P_χ for some $\chi \in X^+(T)$.*

Proof. There is an element w in the Weyl group of T such that $w\chi$ is dominant with respect to $B = TU$ [54; A.10, p.231]. Thus, we may replace χ by $w\chi$ and assume from now on that χ is dominant. Let V be a finite-dimensional, irreducible G-module having highest weight χ. Let $v \in V$ be a highest weight vector, i.e., $v \neq 0$ and $tu \cdot v = \chi(t)v$ for all $tu \in TU$. Let $P_\chi = \{g \in G : g \cdot v = v\}$. Then P_χ contains U. Furthermore, $U_\alpha \subset P_\chi$ if and only if $(\chi, \alpha) \geq 0$. Indeed, if $U_\alpha \subset P_\chi$, then $(\chi, \alpha) \geq 0$ by Theorem 3.2. Also, if $\alpha > 0$, then $U_\alpha \subset P_\chi$ and $U_{-\alpha} \subset P_\chi$ if and only if $(\chi, \alpha) = 0$. We now apply Corollary 3.5 to see that P_χ is generated by all the U_α such that $(\chi, \alpha) \geq 0$ and $\{t \in T : \chi(t) = 1\}$. The group $P = TP_\chi$ contains TU and, so, is parabolic. QED

Corollary 3.7. *Let H be a closed, connected subgroup of G which is normalized by T. Let $\chi \in X(T)$. There is a non-zero $f \in k[G/H]$ such that $r_t \cdot f = \chi(t)f$ for all $t \in T$ if and only if $(\chi, \alpha) \geq 0$ for all $\alpha \in \Phi(T, H)$ and $\chi \mid (T \cap H) = 1$.*

Proof. If such an f exists, then χ must satisfy the stated conditions by Theorem 3.2. Conversely, let V, v, and P_χ be as in the proof of Corollary 3.6. Then, $H \subset P_\chi$ since $U_\alpha \subset P_\chi$ if and only if $(\chi, \alpha) \geq 0$. Let V^* be the dual space of V and let $\mu \in V^*$ be such that $\mu(v) \neq 0$. Put $f(g) = \mu(g \cdot v)$. Then f satisfies the desired conditions. QED

Definition. A subgroup Q of G is said to be *quasi-parabolic* if Q is the stabilizer of a highest weight vector in some finite-dimensional irreducible G-module V. A closed subgroup H of G is called *subparabolic* if there is a quasi-parabolic subgroup Q of G such that $H^o \subset Q$ and $\mathfrak{R}_u H \subset \mathfrak{R}_u Q$.

Lemma 3.8. *Let H be a closed, connected subgroup of G which is normalized by T. Let V be a G-module and let $v \in V$ be such that $H = G_v$. Let $v = \Sigma_\chi v_\chi$ where the v_χ are non-zero T-weight vectors corresponding to distinct T-weights χ. Then, $H = \cap_\chi\{g \in G : g \cdot v_\chi = v_\chi\}$.*

Proof. Obviously, if $g \in G$ and $gv_\chi = v_\chi$ for all χ, then $g \in H$. To prove the converse, we first note that according to Theorem 3.4, $H = H_1 \cdot (T \cap H)^o$ where H_1 is the subgroup of H generated by all the U_α such that $\alpha \in \Phi(T, H)$. Let $S = (T \cap H)^o$. If $s \in S$, then $v = s \cdot v = \Sigma s \cdot v_\chi = \Sigma \chi(s) v_\chi$ so that each $\chi(s) = 1$, i.e., s is in the stabilizer of each v_χ. Next, let $U_\alpha \subset H$ and let $x_\alpha : \mathbf{G}_a \to U_\alpha$ be the canonical homomorphism. According to Lemma 3.1 and its proof, there are vectors $v_{\chi,i} \in V(\chi + i\alpha)$, $i \geq 1$, such that $x_\alpha(c) \cdot v_\chi = v_\chi + \Sigma_i c^i v_{\chi,i}$ for all $c \in k$. Now, $v = x_\alpha(c) \cdot v = \Sigma_\chi x_\alpha(c) v_\chi = \Sigma_\chi (v_\chi + \Sigma_i c^i v_{\chi,i})$. We compare coefficients of the various powers of c to see that each $\Sigma_\chi v_{\chi,i} = 0$ for all i. Since the $v_{\chi,i}$ have different T-weights, each $v_{\chi,i} = 0$ and $x_\alpha(c) v_\chi = v_\chi$. QED

Theorem 3.9. *Let G be a reductive group (not necessarily connected). Let T be a maximal torus in G and let H be a closed subgroup of G which is normalized by T. Then*

H is observable in G if and only if there is an $\chi \in X(T)$ such that H is subparabolic in P_χ.

Proof. According to Corollary 2.2, we may assume that both G and H are connected. We have seen that P_χ is observable in G (Corollary 3.6). Hence, if H is subparabolic, then it is observable in G according to Corollary 2.10. To prove the converse, let V be a finite-dimensional G-module and let $v \in V$ be such that $H = G_v$ (Theorem 1.2). Let $v = \Sigma v_\chi$ where the v_χ's are non-zero T-weight vectors having distinct T-weights. By Lemma 3.8, $H = \cap_\chi \{g \in G : g \cdot v_\chi = v_\chi\}$. Let $\mu = \Sigma\chi$. We now show that H is subparabolic in P_μ. First, recall that (according to Theorem 3.4) $H = H_1 \cdot (T \cap H)^o$ where H_1 is the subgroup of H generated by all the U_α such that $\alpha \in \Phi(T, H)$. If $t \in T \cap H$, then each $\chi(t) = 1$ since $T \cap H$ fixes each v_χ. Thus, $\mu(t) = 1$. If $\alpha \in \Phi(T, H)$, then each $(\chi, \alpha) \geq 0$ by Theorem 3.2 since H fixes v_χ. Thus, $(\mu, \alpha) \geq 0$ and $U_\alpha \subset P_\mu$ by Corollary 3.6. Furthermore, if α is a root in the unipotent radical of H, then some $(\chi, \alpha) > 0$ since $-\alpha$ is not a root of H. Thus, $(\mu, \alpha) > 0$ and α is a root in the unipotent radical of P_μ. QED

Suppose that H is a closed subgroup of G, not necessarily normalized by a maximal torus. We shall show in §7 that H is subparabolic if and only if H is observable in G. We conclude this section with a couple of consequences of the theory just developed.

Lemma 3.10. *Let G be an algebraic group (not necessarily reductive). Let K be a reductive subgroup of G and let H be a closed subgroup of K which contains a maximal torus in K. Then H" is reductive.*

Proof. By Lemma 2.4, $K = K''$ and, so, $H'' \subset K'' = K$. Applying Corollary 2.2 and Exercise 7, §2, we may assume that both H'' and K are connected. Let SU be the radical of H'' where S is a torus and U is a unipotent algebraic group. We shall show that $U = \{e\}$. Let T be a maximal torus of K which is contained in H. Let $\Phi(T,U)$ be the set of roots of U relative to T. Then U is directly spanned by the $\{U_\alpha : \alpha \in \Phi(T,U)\}$ (Lemma 3.3). Let us assume that $\Phi(T,U)$ is not empty. Let $\alpha \in \Phi(T,U)$. Then H'' contains $T \cdot U_\alpha$. But $T \cdot U_\alpha$ is a Borel subgroup of the reductive group G_α which is generated by T, U_α, and $U_{-\alpha}$. According to Lemma 1.1(vi), $(T \cdot U_\alpha)'' = G_\alpha'' = G_\alpha$ and, so, H'' contains G_α. Let $\pi: (H'') \to (H'')/S \cdot U$ be the canonical homomorphism. Since $U_\alpha \subset U \subset \ker\pi$, the derived group of G_α (which is simple) is also contained in $\ker\pi = S \cdot U$. This is impossible since $S \cdot U$ is solvable. QED

To state the last result in this section, we need to recall some definitions and results of [11; Sections 3.8 - 3.13]. A subset Ψ of $\Phi(T,G)$ is *closed* if whenever $\alpha, \beta \in \Psi$ and $\alpha + \beta \in \Phi(T,G)$, then $\alpha + \beta \in \Psi$. It is called *quasi-closed* if the group generated by the U_α, $\alpha \in \Psi$, does not contain any group U_β, $\beta \in \Phi(T,G)$, $\beta \notin \Psi$. A closed subset is always quasi-closed. If Ψ is any subset of $\Phi(T,G)$, the smallest closed (resp. quasi-closed) subset of $\Phi(T,G)$ containing Ψ is called the *closure* (resp. *quasi-closure*) of Ψ. The notions of closed and quasi-closed coincide except when char $k = p = 2,3$. A quasi-closed subset Ψ of $\Phi(T,G)$ determines a subgroup $G(\Psi)$ of G; the subgroup $G(\Psi)$ has a Levi decomposition, i.e. $G(\Psi)$ can be written as a semi-direct product of a reductive subgroup and the unipotent radical of $G(\Psi)$.

Lemma 3.11. *Let G be connected and reductive. Let T be a maximal torus in G. Let H be a closed connected subgroup of G containing T. The following conditions are equivalent.*

(a) *H is epimorphic in G.*

(b) *The quasi-closure of* $\Phi(T,H) \cup -\Phi(T,H)$ *is* $\Phi(T,G)$.

Proof. Suppose that H is epimorphic in G. Let Ψ be the quasi-closure of $\Phi(T,H) \cup -\Phi(T,H)$. Let $G(\Psi) = L \cdot G(\Psi)_u$ be a Levi decomposition. Then $H \subset L$. But $L'' = L$ (Corollary 2.4) so $L = G$. On the other hand, suppose that (b) holds. The group H'' is reductive (by Lemma 3.10) and $\Phi(T,H'')$ is a closed subset of $\Phi(T,G)$ which contains $\Phi(T,H)$ and $-\Phi(T,H)$. By assumption, $\Phi(T,H'') = \Phi(T,G)$ and $H'' = G$. QED

Exercises.

1. Let L be a closed subgroup of G which is normalized by T. Show that the unipotent radical of L is generated by those U_α such that $\alpha \in \Phi(T,L)$ but $-\alpha \notin \Phi(T,L)$.
2. Let G be a connected, reductive algebraic group. Let $H = SU$ where U is a maximal unipotent subgroup and S is a subtorus of T such that $\omega \mid S$ is non-zero for any dominant weight ω. Show that H is epimorphic in G.
3. Let H be the subgroup of $SL_5(k)$ consisting of all those matrices (a_{ij}) with 1's on the diagonal and 0's everywhere else except for a_{13}, a_{14}, a_{24}, and a_{25}. Let T consist of all diagonal matrices in $SL_5(k)$. Describe all those $\chi \in X(T)$ such that $(\chi, \alpha) \geq 0$ for all $\alpha \in \Phi(T, H)$.
4. Let H be a closed, connected subgroup of G which is normalized by T. Show that H is observable in G if and only if there is character $\chi \in X(T)$ such that $(\chi, \alpha) \geq 0$ for all $\alpha \in \Phi(T, H)$, $(\chi, \alpha) > 0$ for all α in the unipotent radical of H and $\chi \mid (T \cap H) = 1$.

5 [82; p. 247]. Let $G = SL_3(k)$. Let T consist of all diagonal elements in G. Let $\Delta = \{\alpha_1, \alpha_2\}$ and let λ be a one-parameter subgroup with $\langle \alpha_i, \lambda \rangle = m_i$, i = 1,2. Let H be the subgroup of G generated by U_α with $\alpha = \alpha_1$ or $\alpha_1 + \alpha_2$ and λ. Show that H is observable in G if $-2 < m_1/m_2 < 1$

6 [76; p.61]. Suppose that char $k = 2$ and let $G = SL_2(k)$. Define a mapping ρ as follows:

$$\rho\begin{pmatrix} a & b \\ c & d \end{pmatrix} = \begin{pmatrix} 1 & ac & bd \\ 0 & a^2 & b^2 \\ 0 & c^2 & d^2 \end{pmatrix}.$$

Show that ρ is a representation of G which is not completely reducible. (Hint: look at the T-weights.)

§4. Finite Generation of $k[G/H]$

We now turn to the question of whether or not the algebra $k[G/H]$ is finitely generated over k. Theorem 1.12 will play a pivotal role here. Indeed, by modifying its proof appropriately, we will be able to give a geometric interpretation for the finite generation of the algebra $k[G/H]$.

Theorem 4.1. *Let H be a closed subgroup of G. If one of the algebras $k[G^o/H^o]$, $k[G/H^o]$, $k[G/H]$, $k[G^o/(H \cap G^o)]$ is finitely generated over k, then all the algebras are.*
Proof. Let $G = \cup_i g_i G^o$ be the coset decomposition of G with respect to G^o. Then $k[G] = \oplus_i k[g_i G^o]$ and, since $H^o \subset G^o$, we have $k[G/H^o] \simeq \oplus_i k[g_i G^o]^{H^o} \simeq \oplus_i k[G^o/H^o]$. Thus, $k[G/H^o]$ is finitely generated over k if and only if $k[G^o/H^o]$ is. Next, if $k[G/H^o]$ is finitely generated over k, then so is $k[G/H]$ since it is the invariants of the finite group H/H^o acting on $k[G/H^o]$.

We now show that if $k[G^o/(H \cap G^o)]$ is finitely generated over k, then so is $k[G^o/H^o]$. For this argument, let us change notation and assume that G is connected and that H is a subgroup of G; thus, we shall show that if $k[G/H]$ is finitely generated over k, then so is $k[G/H^o]$. First, suppose that H is observable in G; then H^o is also observable in G by Corollary 2.2. Let $R = k[G/H]$, $S = k[G/H^o]$, and L be the finite group H/H^o. We note that $S^L = R$. The proof in [10; Proposition 6.15, p.102] shows that S is actually integral over S^L. Furthermore, the quotient field of R is $k(G/H)$ and of S is $k(G/H^o)$ by (5), Theorem 2.1. Now, $k(G/H^o)$ is a finite algebraic extension of $k(G/H)$ according to Theorem 1.10 applied to the mapping $G/H^o \to G/H$ (and using the fact that H/H^o is finite). Thus, by Lemma 1.8, S is the integral closure of R in a finite (algebraic) extension and, so, is finitely generated over k. In general, suppose that H is not observable in G. Let $H = \cup_i h_i H^o$ be the coset decomposition of H with respect to H^o. We shall show that $(H'')^o = (H^o)''$. Let $L = (H^o)''$. The group L is connected by Corollary 2.2; it is normalized by the h_i since they normalize H^o. Therefore, $K = \cup_i h_i L$ is a group which is observable by Corollary 2.2, again. Since K contains H, it contains H''. However, H'' must contain both L and the h_i. Thus, $K = H''$. Now, $k[G/L]$ is finitely generated over k if and only if $k[G/K]$ is according to what was just proved in the observable case. Since $k[G/H] = k[G/K]$ and $k[G/H^o] = k[G/L]$ by Corollary 1.6, we are finished.

To conclude the proof, we show that if $k[G/H]$ is finitely generated over k, then so is $k[G^o/(H \cap G^o)]$. It is enough to show that the restriction mapping from $k[G/H]$ to $k[G^o/(H \cap G^o)]$ is surjective. (This was actually shown in the proof of Corollary 1.3, but we repeat that argument here for convenience.) Consider the coset decomposition of G with respect to G^o, distinguishing between the cosets which contain an element in H and those which do not: $G = \cup_i G^o h_i \cup \cup_j G^o g_j$, where $h_i \in H$ and $G^o g_j \cap H = \varnothing$ for all j. Let $f \in k[G^o/(H \cap G^o)]$ and define $F \in k[G]$ by $F(gh_i) = f(g)$ for $g \in G^o$ and $F = 0$ on each coset $G^o g_j$. We note that if $g, g' \in G^o$ and $h, h' \in H$ satisfy $gh = g'h'$, then $F(gh) = F(g'h') = f(g) = f(g')$ since f is invariant under $H \cap G^o$. It follows that $F \in k[G/H]$ and that F restricts to f. QED

Theorem 4.2. *Let X be an irreducible, normal affine variety. Let X' be a non-empty open subset of X. Then $k[X'] = k[X]$ if and only if $codim(X - X') \geq 2$.*
Proof. (See the appendix to this section.)

Theorem 4.3. *For H an observable subgroup of G, the following conditions are equivalent.*
(a) *There is a finite-dimensional representation $\rho: G \to GL(V)$ and a vector $v \in V$ such that $H = G_v$ and each irreducible component of $cl(G{\cdot}v) - G{\cdot}v$ has codimension ≥ 2 in $cl(G{\cdot}v)$.*
(b) *The algebra $k[G]^H$ is finitely generated over k.*

If (b) holds, let X be the affine variety such that $k[X] = k[G]^H$. *The action of G on* $k[X]$ *by left translation gives an action of G on X. There is a point x in X so that (i)* $G{\cdot}x$ *is open in X, (ii) G/H is isomorphic to* $G{\cdot}x$ *via the mapping* $gH \to g{\cdot}x$, *and (iii) each irreducible component of* $X - G{\cdot}x$ *has codimension* ≥ 2 *in X.*

Proof. Let us first assume that G is connected. Suppose that (a) holds. Let Z be the closure of the orbit $G{\cdot}v$ and let α: $G \to Z$ be given by $\alpha(g) = g{\cdot}v = \rho(g)v$. Let φ: $G/H \to Z$ be given by $\varphi(gH) = g{\cdot}v$. Let X be the normalization of Z in $k(G/H)$. Let π: $X \to Z$ and ψ: $G/H \to X$ be the natural mappings. We have seen (in the second paragraph of the proof of Theorem 1.12) that G/H is isomorphic to the orbit $G{\cdot}x$ where $\psi(eH) = x$. Since $G{\cdot}x$ is invariant under G, so is $X - G{\cdot}x = F$. Since $k[X]$ is integral over $k[Z]$, dim $\pi(F) = \dim F$. Then, we may conclude that $\pi(F) \subset Z - G{\cdot}v$. Indeed, if there were an $a \in F$ with $\pi(a) = g{\cdot}v$ for some $g \in G$, then $\pi(g_1{\cdot}a) = g_1g{\cdot}v$ so $\pi(F)$ would contain all of $G{\cdot}v$; but then $\dim \pi(F) \geq \dim(G{\cdot}v) = \dim Z = \dim X > \dim F$. Therefore, $\pi(X - G{\cdot}x) \subset Z - G{\cdot}v$ and $\dim(X - G{\cdot}x) = \dim \pi(X - G{\cdot}x) = \dim (Z - G{\cdot}v) \leq \dim Z - 2 = \dim X - 2$. We now apply Theorem 4.2 to see that $k[G{\cdot}x] = k[X] = k[G/H]$. This proves statement (b) and the assertions which follow it in case G is connected.

Now, suppose that statement (b) holds (with G connected) and let X be the affine variety such that $k[X] = k[G]^H$. The embedding $k[X] \subset k[G]$ gives rise to a mapping ψ: $G/H \to X$ which is dominant and G-equivariant. Let $x = \psi(eH)$. Then $k[G/H] = k[X] \subset k[G{\cdot}x] \subset k[G/H]$. Thus, $k[G{\cdot}x] = k[X]$ and we may apply Theorem 4.2 always bearing in mind Lemma 1.8. This completes the proof of the equivalence of statements (a) and (b) in the case where G is connected.

In general, consider the coset decomposition of G with respect to G^0, distinguishing between the cosets which contain an element in H and those which do not: $G = \cup_i G^0h_i \cup \cup_j G^0g_j$, where $h_i \in H$, $h_1 = e$ and $G^0g_j \cap H = \emptyset$ for all j. Suppose that (a) holds for G. Then, $\mathrm{cl}(G{\cdot}v) = \mathrm{cl}(G^0{\cdot}v) \cup \cup_j g_j\mathrm{cl}(G^0{\cdot}v)$ with $g_jG^0{\cdot}v \cap \mathrm{cl}(G^0{\cdot}v) = \emptyset$, so each irreducible component of $\mathrm{cl}(G^0.v) - G^0{\cdot}v$ has codimension ≥ 2 in $\mathrm{cl}(G^0{\cdot}v)$ (see Exercise 3). By what was just proved, the algebra $k[G^0/(H \cap G^0)]$ is finitely generated over k. Applying Theorem 4.1, we see that $k[G/H]$ is finitely generated over k, proving (b).

Now, suppose that statement (b) holds and let X be the affine variety such that $k[X] = k[G/H]$. We shall show that X has the properties given in the theorem. There is a morphism φ: $G/H \to X$ whose comorphism, φ^*, is the isomorphism from $k[X]$ to $k[G/H]$. Suppose that $\varphi(eH) = x$; since φ is G-equivariant, we see that $\varphi(gH) = g{\cdot}x$ and, so, $G_x = H$ (by Theorem 2.1(3)) and $G{\cdot}x$ is open and dense in X. For each $f \in k[X]$, we have $\varphi^*(f)(gH) = f(g{\cdot}x)$. Let $X_1 = \mathrm{cl}(G^0{\cdot}x)$. Let ψ: $G^0/(H \cap G^0) \to X_1$ be given by $g(H \cap G^0) \to g{\cdot}x$ and let ψ^* be its comorphism. We first show that $k[X_1] = k[G^0{\cdot}x]$. Let $f' \in k[G^0{\cdot}x]$. Then, $\psi^*(f') \in k[G^0/(H \cap G^0)]$ and, as we saw in the last paragraph of the proof of Theorem 4.1, there is an $F \in k[G/H]$ such that $F \mid (G^0/(H \cap G^0)) = \psi^*(f')$. If $F = \varphi^*F'$ for some $F' \in k[X]$, then $F' \mid G^0{\cdot}x = f'$ since $F'(g^0{\cdot}x) = (\varphi^*F')(g^0) = F(g^0) = (\psi^*f')(g^0) = f'(g^0{\cdot}x)$. Thus, the restriction mapping from $k[G/H] = k[X]$ to $k[G^0{\cdot}x]$ is surjective. Then, the restriction mapping from $k[X_1]$ to $k[G^0{\cdot}x]$ is also surjective and, so, $k[X_1] = k[G^0{\cdot}x]$. A similar argument shows that $k[G^0{\cdot}x] = k[G^0/(H \cap G^0)]$. In fact, let $f \in k[G^0/(H \cap G^0)]$. As noted before, there is an $F \in k[G/H]$ whose restriction to G^0 is f. If $F = \varphi^*(F')$, then a calculation shows that $\psi^*(F' \mid X_1) = f$. Therefore, $k[X_1] = k[G^0{\cdot}x] = k[G^0/(H \cap G^0)]$; Lemma 1.8 and Theorem 4.2 now show that X_1 is normal and $\mathrm{codim}(X_1 - G^0{\cdot}x) \geq 2$. If $G = \cup_i G^0h_i \cup \cup_j G^0g_j$, where $h_i \in H$, $h_1 = $ e and $G^0g_j \cap$

$H = \varnothing$ for all j, then $X = \mathrm{cl}(G{\cdot}x) = X_1 \cup \cup_j g_j X_1$ with $g_j G^o{\cdot}x \cap \mathrm{cl}(G^o{\cdot}x) = \varnothing$ so $\mathrm{codim}(X - G{\cdot}x) \geq 2$. QED

Definition. Let G be a linear algebraic group and let H be an observable subgroup of G. If H satisfies either of the equivalent conditions (a) or (b) in Theorem 4.3, then H is said to satisfy the *codimension 2 condition in G* (or to be a *Grosshans subgroup of G*).

Corollary 4.4. *Let H be a closed subgroup of G. The following statements are equivalent:* (i) *H is a Grosshans subgroup of G,* (ii) *$H \cap G^o$ is a Grosshans subgroup of G^o,* (iii) *H^o is a Grosshans subgroup of G^o.*
Proof. We apply Theorem 4.1 and Corollary 2.2. QED

Corollary 4.5. *Let H and K be observable subgroups of G with $K < H < G$. If K is a Grosshans subgroup of G, then K is a Grosshans subgroup of H.*
Proof. By Corollary 4.4, we may assume that G, H and K are connected. According to Theorem 4.3, there is an affine variety Z on which G acts and a point $z \in Z$ such that (i) $k[Z] = k[G]^K$, (ii) G/K is isomorphic to $G{\cdot}z$, and (iii) the codimension in Z of each irreducible component of $Z - G{\cdot}z$ is ≥ 2. According to Theorem 1.12, there is an affine variety X and a point $x \in X$ so that $X = \mathrm{cl}(G{\cdot}x)$ and G/H is isomorphic to $G{\cdot}x$. The inclusions $k[X] \subset k[G/H] = k[G]^H \subset k[Z]$ give rise to a morphism $f\colon Z \to X$; let $p\colon G/K \to G/H$ be the restriction of f to $G/K = G{\cdot}z$.

Let $W = Z - G{\cdot}z$. We distinguish two cases depending on whether or not $f(W) \cap G{\cdot}x$ is non-empty. First, suppose that there is a $w \in W$ with $f(w) \in G{\cdot}x$. Since W, itself, is G-invariant, we see that $f(W)$ contains G/H. Let W_0 be any irreducible component of W. If W_0 does not dominate X, then there is a non-empty open subset X' of G/H such that $f^{-1}(X') \cap W_0$ is empty. If W_0 dominates X, then there is a non-empty open subset X' in G/H such that for each $gH \in X'$, $f^{-1}(gH)$ has dimension equal to $\dim W_o - \dim G/H$. In either case, we have found a non-empty open subset X' of G/H so that for each $gH \in X'$, we have $\dim (f^{-1}(gH) \cap W_o) \leq \dim W_o - \dim G/H \leq \dim G/K - 2 - \dim G/H = \dim H/K - 2$. Now, $p^{-1}(gH)$ is open in $f^{-1}(gH)$, is isomorphic to H/K, and $f^{-1}(gH) = p^{-1}(gH) \cup (f^{-1}(gH) \cap W)$. But we have just seen that the codimension of $f^{-1}(gH) \cap W$ is ≥ 2 in $f^{-1}(gH)$. We now apply Theorem 4.3.

Finally, suppose that $f(W) \subset X - G{\cdot}x$. Then $f^{-1}(gH) = p^{-1}(gH) = gH/K$, a closed subset of Z. It follows that H/K is affine so K is a Grosshans subgroup of H by Theorem 4.3(b). QED

Corollary 4.6. *If H is reductive, then H is a Grosshans subgroup of G. In fact G/H is affine.*
Proof. Since H is a reductive subgroup of G, the algebra $k[G/H] = k[G]^H$ is finitely generated over k by Theorem A. Let X be as in Theorem 4.3 and let $\varphi\colon G \to X$ be the canonical map. According to Theorem A, again, φ is surjective. Hence, $G/H = X$ is affine. QED

Lemma 4.7. *Let G be solvable. Then any closed subgroup H of G is a Grosshans subgroup, in fact, G/H is affine.*
Proof. This follows immediately from Corollary 2.5. QED

Example 1 [62]. Let $k = \mathbf{C}$ and let G be a connected reductive algebraic group; let G act on its Lie algebra, Lie G, via the adjoint representation. Then, for each $x \in$ Lie G it may be shown that (i) $\mathrm{cl}(G{\cdot}x)$ is a finite union of orbits and (ii) $\dim(G{\cdot}x)$ is even. Thus, for any $x \in \mathrm{Lie}(G)$, we see that $\{g \in G : g{\cdot}x = x\}$ is a Grosshans subgroup of G.

Example 2 [103]. Let G be simply connected and semi-simple. Let $x \in G$ and let $Z_G(x) = \{g \in G : gx = xg\}$ be its centralizer. Then $Z_G(x)$ is a Grosshans subgroup of G. Indeed, it may be shown that properties (i) and (ii) in Example 1 hold in this context.

Example 3 [56]. Let G be a connected, semi-simple algebraic group defined over an algebraic number field. In connection with a number-theoretic problem, the concept of an "absolutely admissible", finite-dimensional, rational representation may be defined and all such representations classified. Given such a representation ρ: $G \to GL(V)$, there is a Zariski-open subset V' of V (the "principal subset") such that for each $x \in V'$, the stabilizer subgroup G_x is Grosshans. The radical of such a G_x is unipotent.

Little is known about the subgroups G_x appearing in Examples 2 and 3. The last three examples may all be viewed as special cases of the following result [85; Theorem, p.7] which will be proved in the appendix to §19.

Lemma 4.8. *Let X be a factorial affine variety on which G acts. There is a dense open subset X' of X such that G_x is a Grosshans subgroup of G for all $x \in X'$.*

Appendix.

In this appendix, we give a proof of Theorem 4.2 based on valuation rings. Although this important concept is covered in standard texts, it is hard to find a reference for Theorem 4.2. Thus, our discussion will be fairly extensive.

Definition. Let K be a field. A subring R of K is called a *valuation ring* if for any non-zero element $u \in K$, either $u \in R$ or $u^{-1} \in R$.

Lemma 1. *Let R be a valuation domain. Let I and J be ideals in R. Then, either $I \subset J$ or $J \subset I$.*

Proof. We may assume that neither ideal is $\{0\}$. Suppose that I is not contained in J and let x be an element in I which is not in J. Let y be any non-zero element in J. Since R is a valuation domain, either x/y or y/x is in R. If $x/y \in R$, then $x = y{\cdot}(x/y) \in J$, a contradiction. Thus, $y/x \in R$ and $y = x{\cdot}(y/x) \in I$. QED

The lemma shows that the ideals in a valuation domain R form a totally ordered set. Now, let M be the unique maximal ideal in R. Then, the set $R - M$ consists of the units in R. For if $x \in R - M$, then the ideal xR is not contained in M. Furthermore, one checks easily that $K - R = \{x \in K^* : x^{-1} \in M\}$.

Definition. Let R be a valuation domain. Let U be the group of units in R. The *value group* of R is K^*/U. If K^*/U is isomorphic to $\mathbf{Z}$, then R is called a *discrete valuation ring*

(denoted by DVR).

Let R be a DVR with, say, $K^*/U = \langle aU \rangle$. We may assume that $a \in M$ (since either a or a^{-1} is). Thus, if $x \in K^*$, we may write x as $a^i u$ where $i \in \mathbf{Z}$ and $u \in U$. If $x \in R$, then $i \geq 0$. Indeed, we may assume that $x \in M$. If $i < 0$, then $a^{-i}x = u$ which contradicts the fact that $a^{-i}x \in M$. It follows that $M = aR$. We also note that if $M = bR$, then $a = bc$, where c is a unit. Thus, for $x \in K^*$ with $x = a^i u$, we also have $x = b^i \cdot$unit.

Theorem 2. *Let R be a valuation domain. If R is a DVR, then R is a principal ideal domain. If R is a principal ideal domain and if R is a proper subset of K, then R is a DVR.*

Proof. Suppose that R is a DVR. We shall use the notation introduced in the paragraph above. Let I be a non-zero ideal in R. Each element in I has the form $a^i u$ where $i \geq 0$ and $u \in U$. We choose $y \in I$ so that i is minimal; we may assume that $y = a^i$. Then $I = yR$. For if $z \in I$, then $z = a^j v$ where $j \geq i$ and $v \in U$. Thus, $z = a^j v = a^i(a^{j-i}v) \in yR$.

Next, suppose that R is a principal ideal domain. In particular, the unique maximal ideal M in R is principal, say $M = aR$, with $a \neq 0$. Let $I = \cap a^i R$ where $i = 1,2, \ldots$. By assumption, there is an element $b \in R$ with $I = bR$. Since I is contained in M, we have $b = az$. Now, $b \in a^i R$ implies that $z \in a^{i-1}R$ for $i = 2,3,\ldots$. Thus, $z \in I$, say $z = bc$ with $c \in R$. Then, $b = az = abc$ and $b(1 - ac) = 0$. Since $a \in M$, we must have $b = 0$, i.e., $I = 0$. Thus, if $x \in R$, there is a non-negative integer i so that x is in $a^i R$ but x is not in $a^{i+1}R$. This means that $x = a^i u$ where $u \in U$. We define a function $v\colon K^* \to \mathbf{Z}$ by $v(x) = i$. Then, one checks that v is a homomorphism and $v(x) = 0$ if and only if $x \in U$. Hence, $K^*/U \simeq \mathbf{Z}$. QED

Definition. Let K be a field. A mapping $v\colon K^* \to \mathbf{Z}$ is called a *discrete valuation* of K if (i) $v(xy) = v(x) + v(y)$ for all $x,y \in K^*$ and (ii) $v(x + y) \geq \min\{v(x), v(y)\}$. Two discrete valuations v_1, v_2 of K are said to be *equivalent* if there are positive integers n_1, n_2 such that $n_1 v_1(x) = n_2 v_2(x)$ for all $x \in K^*$.

Lemma 3. *Let $v\colon K^* \to \mathbf{Z}$ be a discrete valuation. Then*
(1) $v(1) = 0$, (2) $v(x^{-1}) = -v(x)$, and (3) $\{v(x) : x \in K^*\}$ is an ideal in $\mathbf{Z}$.

Proof. First, $v(1) = v(1 \cdot 1) = v(1) + v(1)$ so $v(1) = 0$. Next, $0 = v(1) = v(x \cdot x^{-1}) = v(x) + v(x^{-1})$. Finally, (3) holds since $v(x^m) = mv(x)$. QED

Lemma 4. *Let R be a DVR in K. There is a discrete valuation v with $R = R_v = \{x \in K : v(x) \geq 0\} \sqcup \{0\}$.*

Proof. We have seen that the unique maximal ideal M in R has the form aR for a suitable $a \in R$ and that each $x \in K^*$ may be written as $a^i u$ where $u \in U$. Furthermore, $x \in R$ if and only if $i \geq 0$. Let $v(x) = i$. We shall show that v is a discrete valuation. Property (i) is immediate. As for (ii), let $x_1 = a^i u$, $x_2 = a^j v$ where $i \geq j$ and $u,v \in U$. Then, $x_1 + x_2 = a^j(a^{i-j}u + v)$ where $r = a^{i-j}u + v \in R$. Hence, $v(x_1 + x_2) = j + v(r) \geq j$. QED

Definition. Let k be a subfield of K. A discrete valuation of K is called a *discrete valuation of K/k* if $v(x) = 0$ for all $x \in k^*$.

Theorem 5. *Let v be a (non-zero) discrete valuation of K. Let* $R_v = \{x \in K^* : v(x) \geq 0\} \sqcup \{0\}$ *and* $M_v = \{x \in K^* : v(x) > 0\} \sqcup \{0\}$.
(a) R_v *is a DVR with* M_v *as its (unique) maximal ideal;*
(b) if v_1 *and* v_2 *are equivalent valuations, then* $R_{v1} = R_{v2}$*;*
(c) if v_1 *and* v_2 *are valuations with* $R_{v1} = R_{v2}$*, then* v_1 *is equivalent to* v_2*.*
Proof. Statement (b) is an immediate consequence of the definitions as is the fact in (a) that R_v is a ring with M_v as an ideal. We now complete the proof of statement (a). Let $u \in K^*$, the non-zero elements in the quotient field of R_v. Then, $0 = v(u) + v(u^{-1})$. Thus, either $u \in R_v$ or $u^{-1} \in R_v$ and R_v is a valuation domain. From the preceding equation, we also see that for an element $u \in K^*$, we have that u is a unit in R_v if and only if $v(u) = 0$. Thus, M_v is the unique maximal ideal in R_v. We now show that R_v is a DVR. By Theorem 2, it suffices to show that R_v is a principal ideal domain. Let I be any non-zero ideal in R_v. Then, $I \subset M$. Let $a \in I$ be chosen so that $v(a)$ is as small as possible. Let $b \in I$. Then, $v(b/a) = v(b) - v(a) \geq 0$, so, $b/a \in R_v$ and $b = (b/a)\cdot a \in aR_v$.

Now, we prove statement (c). Let v be any valuation with $R_v = R$. If $x \in K^*$ and $M = aR$, then we have seen that $x = a^i u$ with $u \in U$. Since $v(u) = 0$, we see that $v(x) = iv(a)$. If $R_{v1} = R_{v2} = R_v$, we put $n_1 = v_2(a)$ and $n_2 = v_1(a)$. QED

Taken together, Lemma 4 and Theorem 5 show that there is a one-to-one correspondence between equivalence classes of discrete valuations and DVR's.

Example. Let k be a field and let x be an indeterminate over k. We shall find all the discrete (non-zero) valuations of $k(x)/k$. Let v be a discrete valuation of $k(x)/k$ and let R_v, M_v be as above. Then, either x or x^{-1} is in R_v.

Case 1: *Suppose that* $x \in R_v$*. Then, there is an irreducible polynomial* $p \in k[x]$ *so that* $v(p^e \cdot q/t) = ev(p)$ *where* $q, t \in k[x]$ *with* $(q,p) = (t,p) = 1$*.*
Proof. Since $M_v \cap k[x]$ is a prime ideal in $k[x]$, it has the form $(p) = pk[x]$ where p is irreducible. (If $M_v \cap k[x] = \{0\}$, then all of the non-zero elements in $k[x]$ would be units, $U = k(x)$ and $v = 0$.) If $q \in k[x]$, $q(x) \notin (p)$, then q is a unit and $v(q) = 0$. (One also may check that defining v as in the expression above gives rise to a discrete valuation.)

Case 2: *Suppose that* $x^{-1} \in R_v$*. Then,* $v(p/q) = (\deg q - \deg p)v(1/x)$*.*
Proof. In this case, $M_v \cap k[1/x] = (1/x)$. Also, if $m \geq j$ and $a_m \neq 0$, then $a_m x^m + \ldots + a_j x^j = (1/x)^{-m}(a_m + a_{m-1}(1/x) + \ldots + a_j(1/x)^{m-j})$. The second expression is a unit since it is not in $(1/x)$. Thus, $v(a_m x^m + \ldots + a_j x^j) = -mv(1/x)$. QED

The most important fact concerning DVR's which we shall use is the following result [58; Theorem 104, p.76].

Theorem 6. *Let R be an integrally closed, Noetherian domain.*
(a) Let P be a prime ideal in R having rank 1. Then R_P *is a DVR.*

(b) $R = \cap R_P$ *where P ranges over all prime ideals having rank 1. This representation is locally finite, i.e., any element of R is a unit in all but a finite number of the* R_P*'s.*

Now, let X be an irreducible, normal variety. Let $k[X]$ be the algebra of regular functions on X and let $k(X)$ be the function field of X. Let Z be a closed irreducible subvariety of codimension 1, i.e., a *prime divisor of X*. The local ring of Z consists of all functions in $k(X)$ which are defined on an open subset of Z. This ring is a DVR whose associated valuation we shall denote by v_Z. Indeed, if X is affine, let P be the prime ideal associated to Z. Then, rank $P = 1$ so R_P, the local ring of Z, is a DVR by Theorem 6(a). In general, if X' is an affine, open subset of X which intersects Z, then the local ring of Z is the local ring of $X' \cap Z$.

Definition. Let $f \in k(X)$ and let Z be a closed subvariety of X having codimension 1. If $v_Z(f) > 0$ (resp. < 0), then Z is called a *zero* (resp. *pole*) of f.

Theorem 7. *(a) A function* $f \in k(X)$ *has only finitely many zeros and poles.*
(b) $f \in k[X]$ *if and only if f has no poles.*
(c) Let X' be a non-empty, open subset of X. If X - X' has codimension ≥ 2 *in X, then* $k[X] = k[X']$*. Conversely, if X is affine and if* $k[X'] = k[X]$*, then X - X' has codimension* ≥ 2 *in X.*
Proof. We may assume in proving statements (a) and (b) that X is affine. Let $f = a/b$, where $a,b \in k[X]$. Then a and b are units in all but finitely many R_P (Theorem 6) and, so, statement (a) holds. Also, $f \in R$ if and only if $f \in R_P$ for all P (Theorem 6), i.e, if and only if $v_Z(f) \geq 0$ for all Z. To prove statement (c), let $X - X'$ have codimension ≥ 2. If $f \in k[X']$, then f does not have a pole on X' by (b) so it has no pole on X since the codim $(X - X') \geq 2$. To prove the converse, let X be affine and suppose that some components of $X - X'$ have codimension 1. Let Z_1 be such and let P_1 be its associated prime ideal in $k[X]$. Let $f_1 \in P_1$ generate the maximal ideal in R_{P1}. The function $1/f_1$ has poles at $Z_1, \ldots, Z_m$, say, $v_{Zi}(1/f) = e(i)$. There is a function $f' \in k[X]$ such that $v_{Z1}(f') = 0$ and $v_{Zi}(f') = -e(i)$ for $i \geq 2$ (see Exercise 4). Multiplying $1/f_1$ by f' we obtain a function f such that $v_{Zi}(f) = 0$ for $i \geq 2$, i.e., f has a pole only at Z_1. Then f is in $k[X']$ but not in $k[X]$. QED

We conclude this appendix by describing discrete valuations associated with points in one-dimensional projective space, P^1, and with lines in two-dimensional affine space, $\mathbf{A}^2$. These calculations will be important in Chapter 4.

Let t_0, t_1 be indeterminates. Let F be the field consisting of all rational functions of the form $f(t_0,t_1)/g(t_0,t_1)$ where f and g are homogeneous polynomials in $k[t_0, t_1]$ having the same degree. Let $(\alpha,\beta) \in \mathbf{P}^1$. The point (α,β) gives rise to a discrete valuation v of F by defining $v[(\alpha t_1 - \beta t_0)^m \cdot q_1(t_0,t_1)/q_2(t_0,t_1)] = m$ where q_1 and q_2 are polynomials relatively prime to $(\alpha t_1 - \beta t_0)$. We may identify F with $k(x)$ by the mapping $t_0/t_1 \to x$. In this way, we see that every valuation of $k(x)$ comes from a point in $\mathbf{P}^1$. In fact, $(\alpha,1) \to v[(x - \alpha)^m q_1(x)/q_2(x)] = m$ where $q_i(\alpha) \neq 0$ and $(1,0) \to v(f/g) = \deg g - \deg f$.

Next, consider $\mathbf{A}^2$. Consider a line through the origin, say, $\alpha y - \beta x = 0$. Then $R_P = \{f(x,y)/g(x,y) : (\alpha y - \beta x)$ does not divide $g\}$. The maximal ideal of R_P is generated by $\alpha y - \beta x$. The corresponding valuation v is given by $v[(\alpha y - \beta x)^m \cdot q_1(x,y)/q_2(x,y)] = m$

where each $q_i(x,y)$ is relatively prime to $(\alpha y - \beta x)$. The image of this line in projective space $\mathbf{P}^1$ is the point (α,β). The valuation induced by this point on the field of homogeneous rational functions, $f(x,y)/g(x,y)$ with $\deg f = \deg g$, comes from restricting v to this field.

Exercises.

1. Let H_i, $1 \leq i \leq r$, be observable (resp. Grosshans) subgroups of algebraic groups G_i. Then $H_1 \times \ldots \times H_r$ is an observable (resp. Grosshans) subgroup of $G_1 \times \ldots \times G_r$.

2. Let G and G' be linear algebraic groups and let π: $G' \to G$ be a surjective homomorphism. Let H be a subgroup of G. Then H is observable (resp. Grosshans) in G if and only if $\pi^{-1}(H)$ is an observable (resp. Grosshans) subgroup of G'.

3. With respect to the proof of Theorem 4.3, show that $g_j G^\circ{\cdot}v \cap \mathrm{cl}(G^\circ{\cdot}v) = \varnothing$ and, consequently, that each irreducible component of $\mathrm{cl}(G^\circ.v)$ - $G^\circ{\cdot}v$ has codimension ≥ 2 in $\mathrm{cl}(G^\circ{\cdot}v)$.

4 [66; Proposition 5, p.157]. Let X be a normal, affine variety and let W_i, $1 \leq i \leq n$, be irreducible closed subvarieties having codimension 1. Let m_i, $1 \leq i \leq n$, be non-negative integers. Show that there is a function $f \in k[X]$ such that $v_{W_i}(f) = m_i$ for each i. (Hint: Let P_i be the prime ideal in $k[X]$ belonging to W_i and let $f_i \in P_i$ generate the maximal ideal R_{P_i}. For $j \geq 2$, there is an element $y_j \in P_j$, $y_j \notin P_1$. Then, $y = y_2 \ldots y_n$ does not have a zero at W_1 but does have a zero at each W_j, $j \geq 2$; these zeros can be chosen to have arbitrarily high order by raising y to a suitable power. Let $w_i = f_i^{m_i}$. As was just proved, we may find elements $z_1, \ldots, z_n$ in $k[X]$ such that z_i does not have a zero at W_i but does have a zero of high order (i.e. $> m_j$) at W_j, $j \neq i$. Put $y = z_1 w_1 + \ldots + z_n w_n$. Then, y has the desired property - remember that $v(ab) \geq \min\{v(a), v(b)\}$ and, in fact, is the minimum if $v(a) \neq v(b)$.)

5. Let X be an irreducible, normal affine variety. Let Z_i, $1 \leq i \leq r$, be irreducible subvarieties having codimension 1. Let e_i be integers. Show that there is a function $f \in k(X)$ such that $v_{Z_i}(f) = e_i$.

6*. Lemma 4.8 suggests that the set of Grosshans subgroups may be large in the set of observable subgroups. Can this be made precise?

7*. Prove the Popov-Pommerening conjecture: if H is an observable subgroup of a reductive group G which is normalized by a maximal torus T, then $k[G/H]$ is finitely generated over k.

§5. Maximal Unipotent Subgroups

In this section, we shall show that $k[G/U]$ is finitely generated over k when U is a maximal unipotent subgroup of the linear algebraic group G. In case G is semi-simple, the proof is accomplished by finding explicitly an embedding of G/U which has the codimension 2 property.

Lemma 5.1 [10; Corollary, p.44]. *Let α: $V \to W$ be a dominant morphism of irreducible varieties, where W is normal. Assume that the dimensions of the irreducible components of the fibers of α are constant. Then α is an open map.*

Lemma 5.2. *Let H be a closed subgroup of G. Let Z be any algebraic variety. Then, the mapping $G \times Z \to G/H \times Z$ given by $(g,z) \to (gH,z)$ is open.*

Proof. It is enough to prove this when Z is affine and, indeed, when Z is affine space k^n. In case G is connected, we may apply Lemma 5.1 since $G/H \times k^n$ is normal and the fibers of the mapping are isomorphic to H. In general, $G^o/(H \cap G^o)$ is an open subset of G/H. (Indeed, the homogeneous space G/H may be identified with the orbit of a point x in some projective space [10; Section 6.8, p.98]. Now, let $G = \cup_j g_j G^o$ be the coset decomposition of G with respect to G^o. Then, $G{\cdot}x = \cup_j g_j G^o{\cdot}x$ where the various subsets $g_j G^o{\cdot}x$ are either mutually disjoint or are identical.) Any open subset V in $G \times k^n$ intersects each of the subsets $g_j G^o \times k^n$ in an open set, say V_j. The image of $\ell(g_j^{-1})V_j$ is open in $G^o/(H \cap G^o)$ by what was proved for the connected case. Thus, the image of V_j is open in G/H. QED

Before proving the next theorem, we recall that a variety X is called *complete* if for all varieties Y, the projection mapping of $X \times Y$ to Y sends closed sets to closed sets.

Theorem 5.3. *Let H be a closed subgroup of G such that G/H is complete. Let X be an affine variety on which G acts morphically. Let $x \in X$ and let V (resp. Z) be the closure of the orbit $G{\cdot}x$ (resp. $H{\cdot}x$). Then $V = G{\cdot}Z = \{g{\cdot}z : g \in G, z \in Z\}$.*

Proof. Clearly, $G{\cdot}x \subset G{\cdot}Z \subset V$ so it is enough to show that $G{\cdot}Z$ is closed. Let $S_1 = \{(g,v) \in G \times V : g^{-1}v \in Z\}$. The set S_1 is closed in $G \times V$; furthermore, since Z is H-invariant, we see that $(gh,v) \in S_1$ whenever $(g,v) \in S_1$ and $h \in H$. Let $\pi: G \to G/H$ be the natural mapping and let $\psi = \pi \times \text{id}$ be the mapping from $G \times V$ to $G/H \times V$, given by $\psi(g,v) = (gH,v)$. We note that $\psi^{-1}(\psi(g,v))$ is in S_1 if (g,v) is in S_1. Hence, by Lemma 5.2, the set $S = \psi(S_1) = \{(gH,v) \in G/H \times V : g^{-1}v \in Z\}$ is closed in $G/H \times V$. Since G/H is complete, the projection of S to V is a closed subset of V; but this closed subset is precisely $G{\cdot}Z$. QED

Theorem 5.4. *Let G be a simply connected, semi-simple algebraic group and let U be a maximal unipotent subgroup of G. Then U is a Grosshans subgroup of G.*

Proof. We use the facts about representation theory stated at the beginning of §3. Let $B = TU$ be a Borel subgroup of G and let $\{\omega_1,\ldots,\omega_r\}$ be the fundamental highest weights of G with respect to B. For each $i = 1,\ldots,r$, let V_i be the irreducible G-module having highest weight ω_i and let v_i be a corresponding highest weight vector in V_i. Let $V = V_1 \oplus \ldots \oplus V_r$ and $v = v_1 \oplus \ldots \oplus v_r$. Let Z be the closure of the orbit $G{\cdot}v$. We will show that (a) $U = G_v$ and (b) $\dim(Z - G{\cdot}v) \leq \dim Z - 2$. The finite generation of $k[G]^U$ will then follow from Theorem 4.3.

To prove (a), let $H = \{g \in G : gv_i \in kv_i \text{ for each } i = 1,\ldots,r\}$. We shall show that $H = B$. Obviously, $B \subset H$. If there are elements in H which are not in B, then (using the Bruhat decomposition of G) there is a double coset BsB in H where sT is in the Weyl group of T. Since $s{\cdot}v_i \in kv_i$, we see that $s{\cdot}\omega_i = \omega_i$ for each $i = 1,\ldots,r$. But $\{\omega_1,\ldots,\omega_r\}$ is a **Z**-basis for $X(T)$, so $s \in T$ and $H = B$. Now $U \subset G_v$ since U has no non-trivial character. Let $t \in T \cap G_v$. Then $\omega_i(t) = 1$ for each $i = 1,\ldots,r$. Since $\{\omega_1,\ldots,\omega_r\}$ is a **Z**-basis for $X(T)$, we see that $t = e$ and $U = G_v$.

Finally, we prove (b). Obviously, an element $w \in V$ is in the closure of $B{\cdot}v$ if and only if $w = w_1 \oplus \ldots \oplus w_r$ where $w_i = c_i v_i$, $c_i \in k$. Let $w \in \text{cl}(B{\cdot}v) - B{\cdot}v$. We may assume that $w = v_{m+1} \oplus \ldots \oplus v_r$ where $m \geq 1$. Let $G_w = \{g \in G : g{\cdot}w = w\}$. Obviously, $U \subset G_w$. Let $\alpha = \alpha_1$. The three-dimensional group generated by U_α and $U_{-\alpha}$ is in G_w by Theorem 3.2. Thus, $\dim G{\cdot}w = \dim G - \dim G_w \leq \dim G - \dim U - 2$. Applying Theorem 5.3, we see that $\mathrm{cl}(G{\cdot}v) - G{\cdot}v$ is a union of finitely many orbits $G{\cdot}w$, each of which satisfies the inequality $\dim G{\cdot}w \leq \dim G - \dim U - 2 = \dim G{\cdot}v - 2$. QED

Lemma 5.5. *Let X and Y be normal, irreducible varieties and let $\pi: Y \to X$ be a bijective morphism. If $k[Y]$ is finitely generated over k, then so is $k[X]$.*
Proof. Let $f \in k[Y]$. According to Theorem 1.10, there is a positive integer r, a power of the characteristic of k, so that $f^r \in k(X)$. In fact, we now show that f^r is actually in $k[X]$. For if it is not, then f^r has a pole Z in X (Theorem 7, §4, Appendix) and there is a point $z \in Z$ at which $1/f^r$ is defined and for which $(1/f^r)(z) = 0$. Let $z = \pi(y)$. Then, $1/f^r$ is defined at y and $1 = f^r(y){\cdot}(1/f^r(y)) = f^r(y){\cdot}(1/f^r(z)) = 0$, a contradiction. Now, let $k[Y] = k[f_1, \ldots, f_s]$. For each f_i, choose a positive integer $r(i)$ as above and let $R = k[f_1^{r(1)}, \ldots, f_s^{r(s)}]$. We have seen that R is contained in $k[X]$ and, obviously, $k[Y]$ is integral over R. Thus, $k[Y]$ is a finite R-module; since $k[X]$ is an R-submodule of $k[Y]$, it too is a finite R-module and, so, is finitely generated over k. QED

Theorem 5.6. *Let G be any linear algebraic group and let U be a maximal unipotent subgroup of G. Then U is a Grosshans subgroup of G.*
Proof. The group U is observable in G by Corollary 1.5. We now show that $k[G/U]$ is finitely generated over k in several steps. In so doing, we may assume that G is connected (Corollary 4.4).

First, let G be semi-simple. As we stated at the beginning of §3, there is a simply connected, semi-simple group G_1, having maximal unipotent subgroup U_1, and a surjective homomorphism $\pi: G_1 \to G$ having finite kernel S and satisfying $\pi(U_1) = U$. Then π gives rise to a bijective morphism from G_1/U_1S to G/U. But $k[G_1/U_1]$ is finitely generated over k by Theorem 5.4 and then $k[G_1/U_1S] = k[G_1/U_1]^S$ is finitely generated over k by Theorem A (for finite groups). Thus, $k[G/U]$ is finitely generated over k by Lemma 5.5.

Next, let G be reductive with $G = G_{ss} \times T$ where G_{ss} is semi-simple and T is a torus. Then $U \subset G_{ss}$ and $k[G/U] = k[G_{ss}/U] \otimes k[T]$ is finitely generated by what was just proved. If G is an arbitrary reductive group, there is a reductive group G_1 of the type just studied and a homomorphism $\pi: G_1 \to G$ having finite kernel S. The argument given in the preceding paragraph applies here and shows that $k[G/U]$ is finitely generated over k.

Finally, let G be an arbitrary affine algebraic group. Let $S{\cdot}V$ be the radical of G where S is a torus and V is unipotent. The group $R = G/V$ is reductive with maximal unipotent subgroup U/V. Then $k[G]^U = k[G/U] = k[(G/V)/(U/V)]$ which is finitely generated by the argument given for reductive groups. QED

Embeddings of G/U, where U is maximal unipotent in a semi-simple group G, were completely described in [112] in case char $k = 0$. We shall prove their results in arbitrary characteristic in §17. For the time being, however, let us state the appropriate theorem and give an example. Our notation is as in §3.

Theorem 17.4. *Let G be a reductive algebraic group and let U be a maximal unipotent subgroup of G. Let V be a finite-dimensional G-module and let v be a non-zero element in V such that G_v contains U; let $X = cl(G \cdot v)$. Let $v = \Sigma_i v_i$ where the v_i are non-zero T-weight vectors having distinct T-weights ω_i. Let $ZE = \{\Sigma_i c_i \omega_i : c_i \in Z\}$ and $Q_+E = \{\Sigma_i c_i \omega_i : c_i \in Q, c_i \geq 0\}$. Then $\dim(X - G \cdot v) \leq \dim X - 2$ if and only if $ZE \cap X^+(T) \subset Q_+E$.*

Example. Let $G = SL_3$. Let ω_i, $i = 1,2$, be the fundamental weights of G. Let V_1 (resp. V_2) be the finite-dimensional irreducible G-module having highest weight ω_2 (resp. $\omega_1 + \omega_2$). Let v_1 and v_2 be the corresponding highest weight vectors. Let $V = V_1 \oplus V_2$, $v = v_1 \oplus v_2$, and Z be the closure of the orbit $G \cdot v$. Then $\dim(Z - G \cdot v) = 1$ (Exercise 1).

Example/exercise. Lin Tan has been successful in using Theorem 5.3 to find explicit codimension 2 embeddings of homogeneous spaces G/H in several important cases. We conclude this section with an example which will illustrate his methods. There are many details to be checked; some of these will be left to the reader as an exercise.

Let V be a vector space of dimension 5 over k and let $\{v_1, \dots, v_5\}$ be a basis for V. Let $G = SL_5(k)$. Let B be the subgroup of G consisting of all upper triangular matrices in G, let U be the subgroup of B consisting of all matrices having 1's on the diagonal, and let T be the subgroup of B consisting of diagonal matrices. We define characters χ_i of T for $1 \leq i \leq 5$ as follows: for $t = (t_{ij}) \in T$, let $\chi_i(t) = t_{ii}$. It is known that the simple roots of T relative to B are $\alpha = \chi_1 - \chi_2$, $\beta = \chi_2 - \chi_3$, $\gamma = \chi_3 - \chi_4$, and $\delta = \chi_4 - \chi_5$. Let H be the subgroup of U consisting of all those matrices (a_{ij}) with 1's on the diagonal, a_{13}, a_{14}, a_{24}, a_{25} arbitrary elements in k, and 0's elsewhere. The subgroup H, being unipotent, is observable in G. We wish to show that H is a Grosshans subgroup of G, i.e., $k[G/H]$ is finitely generated over k [109].

Let ΛV be the exterior algebra over V and let $\Lambda^i V$ be the subspace of ΛV consisting of those elements having step i. The group G acts naturally on ΛV and $\Lambda^i V$. Let $v = (v_1 \wedge v_3) \oplus (v_2 \wedge v_5) \oplus (v_1 \wedge v_2 \wedge v_3 \wedge v_4) \oplus (v_1 \wedge v_2 \wedge v_4 \wedge v_5) \in \Lambda^2 V \oplus \Lambda^2 V \oplus \Lambda^4 V \oplus \Lambda^4 V$. The T-weights of the various components of v are of the form $w \cdot \omega$ where w is in the Weyl group of T and ω is a highest weight with respect to B. Furthermore, these weights form a basis of $X(T)$.

Let W be a subspace of V having basis $\{w_1, \dots, w_r\}$. It is known that $g \cdot W \subset W$ if and only if $g \cdot (w_1 \wedge \dots \wedge w_r) = c(w_1 \wedge \dots \wedge w_r)$ for some $c \in k^*$. Using this, it may be shown that the stabilizer in G of v is H. We will now examine $X = cl(G \cdot v)$ by first looking at $cl(B \cdot v)$ and then using Theorem 5.3. In fact, we show that $cl(B \cdot v) - B \cdot v$ is the union of finitely many subvarieties S where $\dim(G \cdot S) \leq \dim(G/H) - 2$. This will suffice to prove the result.

The B-orbit of v consists of all vectors having the form (*)

$$\begin{aligned} &av_1 \wedge v_3 + a_1 v_1 \wedge v_2 \oplus dv_2 \wedge v_5 + d_1 v_1 \wedge v_5 + d_2 v_1 \wedge v_4 \\ &\quad + d_3 v_2 \wedge v_4 + d_4 v_2 \wedge v_3 + d_5 v_1 \wedge v_3 + d_6 v_1 \wedge v_2 \\ &\oplus bv_1 \wedge v_2 \wedge v_3 \wedge v_4 \\ &\oplus cv_1 \wedge v_2 \wedge v_4 \wedge v_5 + c_1 v_1 \wedge v_2 \wedge v_3 \wedge v_5 \\ &\quad + c_2 v_1 \wedge v_2 \wedge v_3 \wedge v_4 \end{aligned}$$

where $abcd \neq 0$ and (**) $dd_2 = d_1 d_3$, $d_1 d_4 = dd_5$, $d_1 c_2 = c_1 d_2 - cd_5$, $dc_2 = c_1 d_3 - cd_4$. In fact, let $(b_{ij}) \in B$. We calculate $(b_{ij}) \cdot v$ and see that $a = b_{11} b_{33}$, $a_1 = b_{11} b_{23}$, $d = b_{22} b_{55}$,

$d_1 = b_{12}b_{55}$, $d_2 = b_{12}b_{45}$, $d_3 = b_{22}b_{45}$, $d_4 = b_{22}b_{35}$, $d_5 = b_{12}b_{35}$, $d_6 = -b_{15}b_{22}$, $b = b_{11}b_{22}b_{33}b_{44}$, $c = b_{11}b_{22}b_{44}b_{55}$, $c_1 = b_{11}b_{22}b_{34}b_{55}$, $c_2 = b_{11}b_{22}b_{34}b_{45} - b_{11}b_{22}b_{44}b_{35}$. (In looking at the B-orbit, we may take $b_{13} = b_{14} = b_{24} = b_{25} = 0$ since H stabilizes v.) Then, given a vector w of the form (*) and satisfying equations (**) with $abcd \neq 0$, we may solve the equations $(b_{ij})\cdot v = w$ involving all a_i, b, c_i, d_i except possibly c_2, d_2, d_5. But those equalities must also hold since the entries of both w and $(b_{ij})\cdot v$ satisfy (**) and $d \neq 0$. To examine the boundary of $B\cdot v$, i.e., $\mathrm{cl}(B\cdot v) - B\cdot v$, we distinguish two cases depending on whether or not $d = 0$.

$\mathbf{d \neq 0}$. The boundary consists of all the points of the form (*), satisfying equations (**), but with $abcd = 0$. For given such a point w, we can find points in $B\cdot v$ with all the same entries as w except for a, b, c being arbitrary non-zero elements in k and c_2 satisfying the equation $c_2 = (c_1d_3 - cd_4)/d$. Any polynomial vanishing at all such points must also vanish at w.

Suppose first that $a = 0$, but $a_1bcd \neq 0$. Let $v' = (v_1 \wedge v_2) \oplus (v_2 \wedge v_5) \oplus (v_1 \wedge v_2 \wedge v_3 \wedge v_4) \oplus (v_1 \wedge v_2 \wedge v_4 \wedge v_5)$. Then, reasoning as before, we see that $B\cdot v'$ consists of all points of the form (*), satisfying equations (**), except that $a = 0$ and $a_1bcd \neq 0$; also, the boundary of $B\cdot v'$ consists of all such points except $a_1bcd = 0$. A calculation shows that the stabilizer in G of v' contains H and $U_{-\alpha}$, U_β, and $U_{-\gamma}$. Consequently, $\dim G\cdot v' \leq \dim(G/H) - 3$. A similar argument applies when $b = 0$. Now, suppose that $c = 0$. Let S consist of all vectors of the form $(v_1 \wedge v_3) \oplus (v_2 \wedge v_5) \oplus (v_1 \wedge v_2 \wedge v_3 \wedge v_4) \oplus x(v_1 \wedge v_2 \wedge v_3 \wedge v_5)$. (We note that the characters of T corresponding to the various components of such vectors are no longer independent.) If w is in the boundary of $B\cdot v$ with $c = 0$, then (reasoning as before) we may find a vector w' in S, with $x \neq 0$, so that w is in $\mathrm{cl}(B\cdot w')$. We consider the mapping $G \times S \to \mathrm{cl}(G\cdot v)$. The fiber above a point has dimension ≥ 7. (Any element in G which fixes $v_1 \wedge v_3$, $v_2 \wedge v_4$, and $v_2 \wedge v_5$ fixes each point in S.) Hence, the image has dimension $\leq \dim G + \dim S - 7 = 18 = \dim(G/H) - 2$, as desired.

$\mathbf{d = 0, d_1 \neq 0}$. The equations (**) now become $d_1d_3 = 0$, $d_1d_4 = 0$, $d_1c_2 = c_1d_2 - cd_5$, $c_1d_3 = cd_4$. Then $d = d_3 = d_4 = 0$ and $d_1c_2 = c_1d_2 - cd_5$. Let S consist of all vectors of the form $(v_1 \wedge v_3) \oplus x(v_1 \wedge v_5) \oplus (v_1 \wedge v_2 \wedge v_3 \wedge v_4) \oplus (v_1 \wedge v_2 \wedge v_4 \wedge v_5)$ and proceed as in the case $c = 0$ above.

$\mathbf{d = d_1 = 0, c \neq 0}$. The equations (**) now become $c_1d_2 = cd_5$ and $c_1d_3 = cd_4$. Let S consist of all vectors of the form $(v_1 \wedge v_3) \oplus x(v_2 \wedge v_4) \oplus (v_1 \wedge v_2 \wedge v_3 \wedge v_4) \oplus (v_1 \wedge v_2 \wedge v_4 \wedge v_5)$ and proceed as before.

$\mathbf{d = d_1 = c = 0, c_1 \neq 0}$. The equations (**) now become $c_1d_2 = c_1d_3 = 0$. Let $S = \mathrm{cl}(B\cdot v')$ where $v' = (v_1 \wedge v_3) \oplus (v_2 \wedge v_3) \oplus (v_1 \wedge v_2 \wedge v_3 \wedge v_4) \oplus (v_1 \wedge v_2 \wedge v_3 \wedge v_5)$. (We note that the characters of T corresponding to the various components of v' are now linearly independent.)

$\mathbf{d = d_1 = c = c_1 = 0}$. All the equations (**) are then satisfied and we let S consist of all vectors of the form $(v_1 \wedge v_3) \oplus [x(v_1 \wedge v_3) + y(v_2 \wedge v_4)] \oplus (v_1 \wedge v_2 \wedge v_3 \wedge v_4) \oplus z(v_1 \wedge v_2 \wedge v_3 \wedge v_4)$.

Exercises.

1. Check the facts given in the Example following Theorem 5.7.
2. Let $G = SL_2(k)$ and let U be a maximal unipotent subgroup of G. Show that the condition in Theorem 17.4 always holds, i.e., $\mathbf{Z}E \cap X^+T \subset \mathbf{Q}_+E$.
3. Let $B = TU$ be a Borel subgroup of G. Let ω be a weight of $X(T)$ and define $E(\iota\omega) = \{f \in k[G/U] : r_t f = \omega(t) f \text{ for all } t \in T\}$. Show that $E(\iota\omega)$ has finite dimension. (Hint: reduce to the case where G is reductive.)

Bibliographical note

The main sources for the material on observable groups in this chapter are [3], [4], [82], and [106]. The material in §3 is largely based on the exposition in [82]. The basic facts about Grosshans subgroups may be found in [38] and [83].

Chapter Two

The Transfer Principle

Introduction. In §6, we introduce the concept of an induced module and prove some basic properties. Induced modules are related to observable subgroups in §7 and to finite generation questions in §8 and §9. In §7, we prove Sukhanov's group-theoretic characterization of observable subgroups along with several other results. In §8, we show that if G is not reductive, then there are always finitely generated, commutative k-algebras A on which G acts rationally, but for which the algebra of invariants, A^G, is not finitely generated over k. Our approach to obtaining affirmative answers to the finite-generation question is dictated by the following result (§9). Let H be a closed subgroup of G and let W be a G-module; then $(k[G/H] \otimes W)^G \simeq W^H$, where the action of G on $k[G/H]$ is by left translation. Since the nineteenth century (and, especially the work of Felix Klein), this has been called the "adjunction principle". Recently, A. Borel has reformulated it (as below) and introduced the name "transfer principle". Roughly speaking, it allows information on $k[G/H]$ to be transferred to W^H. For example, suppose that G is reductive and that $k[G/H]$ is finitely generated. Let $W = A$ be a finitely generated, commutative k-algebra on which G acts rationally. Then using the transfer principle and Theorem A, we see that A^H is finitely generated. The most important instance of this occurs when $H = U$ is a maximal unipotent subgroup of a reductive group G (§9). In Sections 10 and 11, we look at other important (classical) examples. First, in the study of binary forms (§10), H is taken to be a maximal unipotent subgroup of $SL(2,\mathbf{C})$. The transfer theorem in this context was proved by M. Roberts in 1871 and describes the relationship between "covariants", i.e. the algebra $(\mathbf{C}[G/H] \otimes A)^G$, and "semi-invariants", the algebra A^H [33; p.125]. The examples studied in §11 are first placed in the context of multiplicity-free actions. They reflect a rich heritage. By choosing H to be an orthogonal subgroup of $SL(3,\mathbf{R})$, Felix Klein was able to give a unified treatment through projective geometry of euclidean and non-euclidean geometries [59; p.179]. H.W. Turnbull proved the transfer theorem when H is an orthogonal or affine subgroup of $SL(n,\mathbf{C})$. In the former case he described transfer theorems as being of "high interest in metrical geometry" [110; p.324]. In these same contexts, H. Weyl proved the transfer theorem using the symbolic method and then said that this is a "triumphant attainment of the symbolic method" [116; p.257]. Finally, D.E. Littlewood proved the transfer theorem when H is orthogonal or symplectic and used it to construct invariants of these groups; he called the result a "fundamental theorem" [68; p.391].

§6. Induced Modules

Throughout this section, let H be a closed subgroup of G. Let V be an H-module. The given action of H on V together with the action of H on $k[G]$ by right translation gives an action of H on $k[G] \otimes V$. The group G acts trivially on V and by left translation on $k[G]$. The actions of H and G on $k[G] \otimes V$ commute so G acts on $(k[G] \otimes V)^H$. The G-module so defined is denoted by $\mathrm{ind}_H{}^G V$ and is called the *induced module of V from*

H to G. The algebra $k[G]^H$ acts on $k[G] \otimes V$ by $a \cdot \Sigma_i f_i \otimes v_i = \Sigma_i af_i \otimes v_i$. This action also commutes with the action of H on $k[G] \otimes V$. Hence, $(k[G] \otimes V)^H$ is a $k[G]^H$-module. Our purpose in this section is to prove some basic properties of induced representations. They shall provide the basis for our future study of the finite-generation question. One more bit of notation is in order here: if W is a G-module, then when we consider W as an H-module we write $W|_H$.

Example. Let $V = k$ with the trivial action of H. Then $\text{ind}_H{}^G V = k[G]^H$ and the action by G is by left translation. In particular, if $H = \{e\}$, then $\text{ind}_{\{e\}}{}^G k = k[G]$.

It will often be convenient to use another definition of induced representations. Let V be an H-module with respect to the mapping $H \times V \to V$, $(h,v) \to h*v$. Let $\{v_i\}$ be a basis for V. A *morphism* from G to V is any mapping f: $G \to V$ of the form $f(g) = \Sigma_i f_i(g)v_i$ where each $f_i \in k[G]$ and where only finitely many of the f_i's are non-zero. We shall denote the vector space of all such morphisms by Mor(G,V). The vector space $k[G] \otimes V$ may be identified with Mor(G,V) by associating to $\Sigma_i f_i \otimes v_i$ the morphism $g \to \Sigma_i f_i(g)v_i$. The group H acts on Mor(G,V) by $(h*f)(g) = h*f(gh)$; it acts on $k[G] \otimes V$ by right translation on $k[G]$ and by the given representation on V. The isomorphism is H-equivariant with respect to these H-actions. Thus, $\text{ind}_H{}^G V$ may be identified with the subspace of Mor(G,V) consisting of those f such that $h*f(gh) = f(g)$. We note that the action of G on Mor(G,V) and on $\text{ind}_H{}^G V$ is given by $(g \cdot f)(x) = f(g^{-1}x)$. The action of $k[G]^H$ on $\text{ind}_H{}^G V$ is given by $(a \cdot f)(g) = a(g)f(g)$.

Let E_V: $\text{ind}_H{}^G V \to V$ be given by $E_V(f) = f(e)$ where, as usual, e is the identity element in G. Then E_V is H-equivariant since $E_V(h \cdot f) = (h \cdot f)(e) = f(h^{-1}) = h*f(h^{-1}h) = h*f(e) = h*E_V(f)$.

(P1) (universal mapping property). *Let V (resp. W) be a rational H-module (resp. G-module). Let φ: $W \to V$ be an H-homomorphism. Then, there is a unique G-homomorphism φ_1: $W \to \text{ind}_H{}^G V$ such that $E_V \circ \varphi_1 = \varphi$.*

Proof. For $w \in W$, we define $\varphi_1(w)$: $G \to V$ by $\varphi_1(w)(g) = \varphi(g^{-1} \cdot w)$. Since the action of G on W is locally finite, there are elements w_i in W and functions f_i in $k[G]$ so that $g^{-1} \cdot w = \Sigma_i f_i(g)w_i$. Thus, we see that $\varphi_1(w) \in \text{Mor}(G,V)$. That φ_1 has the desired properties follows immediately. To prove uniqueness, suppose that φ_2: $W \to \text{ind}_H{}^G V$ is another G-homomorphism having the desired properties; let $\varphi' = \varphi_1 - \varphi_2$. Then $E_V \circ \varphi' = 0$ and we wish to show that $\varphi' = 0$. Since φ' is a G-homomorphism, we have $\varphi'(w)(g) = \varphi'(w)(ge) = (g^{-1} \cdot \varphi'(w))(e) = \varphi'(g^{-1} \cdot w)(e) = E_V(\varphi'(g^{-1} \cdot w)) = 0$. QED

(P2) (the universal mapping property characterizes $\text{ind}_H{}^G$). *Suppose that V' is a G-module and that there is an H-equivariant linear mapping E': $V' \to V$ so that the following holds: for any G-module W and any H-homomorphism φ: $W \to V$, there is a unique G-homomorphism ψ: $W \to V'$ so that $E' \circ \psi = \varphi$. Then V' is isomorphic as a G-module to $\text{ind}_H{}^G V$.*

Proof. By assumption, there is a G-homomorphism ψ: $\text{ind}_H{}^G V \to V'$ such that $E' \circ \psi = E_V$. Also, by the universal mapping property, there is a G-homomorphism ψ': $V' \to \text{ind}_H{}^G V$ such that $E_V \circ \psi' = E'$. Then, $E_V \circ (\psi' \circ \psi) = E' \circ \psi = E_V$; by uniqueness, $\psi' \circ \psi$ is the identity on $\mathrm{ind}_H{}^G V$. Similarly, $\psi \circ \psi'$ is the identity on V'. QED

(P3) (reciprocity of induction). *Let V (resp. W) be a rational H-module (resp. G-module). Then $Hom_G(W, ind_H{}^G V) \simeq Hom_H(W|_H, V)$.*

Proof. This is just a restatement of the universal mapping property, the isomorphism being given by $\varphi \in \mathrm{Hom}_G(W, \mathrm{ind}_H{}^G V) \to E_V \circ \varphi$. QED

(P4) (transitivity of induction). *Let $H \subset L \subset G$ be closed subgroups of G. Let V be an H-module. Then $ind_L{}^G(ind_H{}^L V) = ind_H{}^G V$. The algebra $k[G]^L$ acts on $ind_L{}^G(ind_H{}^L V)$ and on $ind_H{}^G V$ since $k[G]^L \subset k[G]^H$. The isomorphism is $k[G]^L$-equivariant.*

Proof. We shall check that the conditions of (P2) hold for $V' = \mathrm{ind}_L{}^G(\mathrm{ind}_H{}^L V)$ and $E' = E_L \circ E_G$ where E_L: $\mathrm{ind}_H{}^L V \to V$ and E_G: $\mathrm{ind}_L{}^G(\mathrm{ind}_H{}^L V) \to \mathrm{ind}_H{}^L V$ are the canonical mappings. To that end, let W be a G-module and let φ: $W \to V$ be an H-equivariant morphism. Since W is an L-module, there is an L-equivariant mapping φ_1: $W \to \mathrm{ind}_H{}^L V$ such that $E_L \circ \varphi_1 = \varphi$. Similarly, there is a G-equivariant mapping φ_2: $W \to \mathrm{ind}_L{}^G(\mathrm{ind}_H{}^L V)$ such that $E_G \circ \varphi_2 = \varphi_1$. Then, $E' \circ \varphi_2 = (E_L \circ E_G) \circ \varphi_2 = E_L \circ \varphi_1 = \varphi$. To prove that the mapping is uniquely determined, suppose that $E_L \circ E_G \circ \varphi = 0$ where φ: $W \to \mathrm{ind}_L{}^G(\mathrm{ind}_H{}^L V)$ is G-equivariant. Then $E_L \circ (E_G \circ \varphi) = 0$ implies that $E_G \circ \varphi = 0$ and, so, $\varphi = 0$.

Next, we show that the isomorphism is $k[G]^L$-equivariant, i.e., if ψ: $\mathrm{ind}_H{}^G V \to \mathrm{ind}_L{}^G(\mathrm{ind}_H{}^L V)$ is the isomorphism and if $a \in k[G]^L$, then $\psi(a{\cdot}f) = a{\cdot}\psi(f)$. First, we need to recall the definition of the mapping ψ. To do this, we retrace the argument just given replacing W by $\mathrm{ind}_H{}^G V$. Let ψ_1: $\mathrm{ind}_H{}^G V \to \mathrm{ind}_H{}^L V$ be the L-morphism such that $E_L \circ \psi_1 = E_V$. By definition (see the first line of the proof of (P1)), $\psi_1(f)(\ell) = E_V(\ell^{-1}{\cdot}f) = (\ell^{-1}{\cdot}f)(e) = f(\ell)$ for all $\ell \in L$. Similarly, there is a mapping ψ_2: $\mathrm{ind}_H{}^G V \to \mathrm{ind}_L{}^G(\mathrm{ind}_H{}^L V)$ such that $E_G \circ \psi_2 = \psi_1$ and, $(\psi_2 f)(g) = \psi_1(g^{-1}{\cdot}f)$ for all $g \in G$. Then, $\psi = \psi_2$ and $\psi_2(a{\cdot}f)(g)(\ell) = \psi_1(g^{-1}(a{\cdot}f))(\ell) = g^{-1}(a{\cdot}f)(\ell) = (a{\cdot}f)(g\ell) = a(g\ell)f(g\ell) = a(g)f(g\ell)$. Also, $(a{\cdot}\psi_2 f)(g)(\ell) = (a(g)(\psi_2 f)(g))(\ell) = a(g)\psi_1(g^{-1}{\cdot}f)(\ell) = a(g)f(g\ell)$. QED

(P5) (tensor identity). *Let W be a G-module. Then $ind_H{}^G V \otimes W \simeq ind_H{}^G(V \otimes W|_H)$. The isomorphism is G-equivariant where the action of G on the left-hand side is by the given action on W and the usual action on $ind_H{}^G V$. The isomorphism is also $k[G]^H$-equivariant where $k[G]^H$ acts on the left-hand side by $a{\cdot}\Sigma_i f_i \otimes w_i = \Sigma_i af_i \otimes w_i$.*

Proof. Let $f \in \mathrm{ind}_H{}^G V$, $w \in W$ and define $\psi(f \otimes w)(g) = f(g) \otimes g^{-1}w$. Then $\psi(f \otimes w) \in \mathrm{ind}_H{}^G(V \otimes W|_H)$. Indeed, $\psi(f \otimes w)$ is a morphism since f is and G acts rationally on W. Furthermore, for $h \in H$ we have $h{\cdot}\psi(f \otimes w)(gh) = h{\cdot}(f(gh) \otimes h^{-1}g^{-1}w) = hf(gh) \otimes g^{-1}w = f(g) \otimes g^{-1}w = \psi(f \otimes w)(g)$.

The mapping ψ is injective. For let $\{w_i\}$ be a basis for W. If $\psi(\Sigma_i f_i \otimes w_i) = 0$, then $\Sigma_i f_i(g) \otimes g^{-1}w_i = 0$ for all $g \in G$. Since the vectors $g^{-1}w_i$ are linearly independent, each $f_i(g) = 0$ so each $f_i = 0$. To show that ψ is surjective, let $F \in \mathrm{ind}_H{}^G(V \otimes W|_H)$. Then $F(g) = \Sigma_i a_i(g) \otimes w_i$ where the a_i are morphisms from G to V and we may assume that the set $\{w_i\}$ is finite and spans a G-invariant subspace. Now, $g{\cdot}w_i = \Sigma_j b_{ij}(g)w_j$ and, so, $w_i = \Sigma_j b_{ij}(g)g^{-1}w_j$. Hence, $F(g) = \Sigma_i f_i(g) \otimes g^{-1}w_i$ where $f_i \in \mathrm{Mor}(G, V)$. We need to prove that $h{\cdot}f_i(gh) = f_i(g)$ for then $\psi(\Sigma_i f_i \otimes w_i) = F$. But $h{\cdot}F(gh) = F(g)$ implies that $h{\cdot}\Sigma_i f_i(gh) \otimes h^{-1}g^{-1}w_i = \Sigma_i f_i(g) \otimes g^{-1}w_i$. Hence, $\Sigma_i hf_i(gh)$

$\otimes\ g^{-1}w_i = \Sigma_i f_i(g) \otimes g^{-1}w_i$ and $hf_i(gh) = f_i(g)$.

The mapping ψ is G-equivariant. For let $x \in G$. Then, $\psi(xf \otimes xw)(g) = (xf)(g) \otimes g^{-1}\cdot(x\cdot w) = f(x^{-1}g) \otimes g^{-1}x\cdot w$. Also, $x\cdot\psi(f \otimes w)(g) = \psi(f \otimes w)(x^{-1}g) = f(x^{-1}g) \otimes g^{-1}x\cdot w$. Finally, we show that ψ is $k[G]^H$-equivariant. Let $a \in k[G]^H$. Then $\psi(af \otimes w)(g) = (af)(g) \otimes g^{-1}w = a(g)f(g) \otimes g^{-1}w$ while $a\cdot\psi(f \otimes w)(g) = a(g)(\psi(f \otimes w)(g)) = a(g)f(g) \otimes g^{-1}w$. QED

(P6) *Let W be a G-module. Then $k[G]^H \otimes W \simeq ind_H{}^G(W|_H)$ where $k[G]^H$ acts on the left-hand side by $a\cdot(f \otimes w) = af \otimes w$ and G acts on the left-hand side by $g\cdot(f \otimes w) = \ell_g f \otimes g\cdot w$.*

Proof. In (P5), let $V = k$ with the trivial action by H. QED

(P7) *Let W be a G-module. Then $ind_G{}^G W \simeq W$ as G-modules.*

Proof. In (P6), let $H = G$. QED

(P8) (the transfer principle). *Let W be a G-module.*

(a) *$(k[G] \otimes W)^G \simeq W$ as G-modules where, on the left-hand side, $(k[G] \otimes W)^G = \{\Sigma_i f_i \otimes w_i \in k[G] \otimes W\colon \Sigma_i \ell_g f_i \otimes g\cdot w_i = \Sigma_i f_i \otimes w_i$ for all $g \in G\}$ and G acts by right translation on $k[G]$;*

(b) *$(k[G]^H \otimes W)^G \simeq W^H$.*

Proof. Let ϵ be the automorphism from $k[G]$ to $k[G]$ given by $(\epsilon f)(x) = f(x^{-1})$; then $\epsilon \circ r_g = \ell_g \circ \epsilon$ for all $g \in G$. Since we can use ϵ to switch left and right translation, we see from (P7) that $(k[G] \otimes W)^G \simeq W$ as G-modules where G acts by right translation on $k[G]$. It follows that the subspaces consisting of the H-fixed points are isomorphic. But the actions of G and H commute so that $((k[G] \otimes W)^G)^H = ((k[G] \otimes W)^H)^G = (k[G/H] \otimes W)^G$. QED

(P9) *The induction functor is left exact, i.e., if V_1, V_2, and V_3 are H-modules and if there are H-homomorphisms so that $0 \to V_1 \to V_2 \to V_3$ is exact, then $0 \to ind_H{}^G V_1 \to ind_H{}^G V_2 \to ind_H{}^G V_3$ is also exact.*

Proof. (Exercise 2).

(P10) *The quotient space G/H is affine if and only if the induction functor is exact.*

Proof. This rather long proof is in the appendix which follows.

Appendix.

To prove that induction is exact when G/H is affine, we shall follow the argument given in [19]. That proof relies on properties of flat modules (for which we shall use [71] as a reference) and, most importantly, on the fact that $k[G]$ is faithfully flat over $k[G/H]$. Throughout this appendix, let H be a closed subgroup of G and let V be an H-module.

Note. To show that induction is exact is equivalent to proving that $R^1(\mathrm{ind}_H{}^G V_1) = 0$. Actually, something more is true, namely, that this is equivalent to $H^1(H, k[G] \otimes V) = 0$ for every H-module V [19, p.7].

Definition. A G-module Q is *rationally injective* if for any G-modules M,N with $M \subset N$ and any G-homomorphism $\alpha: M \to Q$, there is a G-homomorphism $\beta: N \to Q$ such that $\beta = \alpha$ on M.

Lemma 1. (a) *A G-direct summand of a rationally injective G-module is rationally injective.*
(b) *Let Q be a rationally injective G-module and let $0 \to Q \to V \to W \to 0$ be an exact sequence of G-modules and G-homomorphisms. Then, there is a G-submodule Q' of V so that $V = Q \oplus Q'$ and the mapping $Q' \to W$ is an isomorphism. Furthermore, we have the exact sequence $0 \to Q^G \to V^G \to W^G \to 0$.*
(c) *Let H be a closed subgroup of G and let V be a rationally injective H-module. Then $\mathrm{ind}_H{}^G V$ is a rationally injective G-module.*
(d) *$k[G]$ is a rationally injective G-module.*
Proof. To prove (a), let W be a rationally injective G-module and let $W = Q \oplus Q'$ where Q and Q' are G-submodules. Let M, N, and $\alpha: M \to Q$ be as in the definition. Then α gives a G-homomorphism $\alpha_1: M \to W$ by $\alpha_1(m) = (\alpha(m),0)$. Since W is rationally injective, there is a G-homomorphism $\beta_1: N \to W$ so that $\beta_1 = \alpha_1$ on M. Let $\beta = p \circ \beta_1$ where p is the projection of W on Q.

To prove (b), let $\alpha: Q \to Q$ be the identity map and let $\beta: V \to Q$ be as in the definition. Let Q' be the kernel of β. Then one checks directly that Q' has the desired properties. The rest of the statement follows at once.

Let M,N be G-modules and let $\alpha: M \to \mathrm{ind}_H{}^G V$ be as in the definition. The mapping $E_V \circ \alpha$ is a H-homomorphism from M to V. Since V is rationally injective, there is an H-homomorphism $\beta: N \to V$ whose restriction to M is $E_V \circ \alpha$. According to (P1), then, there is a G-homomorphism $\beta_1: N \to \mathrm{ind}_H{}^G V$ so that $E_V \circ \beta_1 = \beta = E_V \circ \alpha$ on M. Then (as in the uniqueness argument in (P1)), $\beta_1 = \alpha$ on M. To prove statement (d), we simply note that $k[G] = \mathrm{ind}_{\{e\}}{}^G k$; any $\{e\}$-module is rationally injective since (with the notation of the definition) a k-basis of M can be extended to N. QED

Theorem 2 (Hochschild). *Let Q be a rationally injective G-module. Let W be any rational G-module. Then $W \otimes Q$ is a rationally injective G-module.*
Proof. Let G act on $k[G]$ by left translation. According to (P6), the G-module $W \otimes Q \otimes k[G]$ is $\mathrm{ind}_{\{e\}}{}^G(W \otimes Q)$ and is rationally injective by Lemma 1(c). On the other hand, we have an exact sequence $0 \to Q \to Q \otimes k[G] \to Q_1 \to 0$ where $q \to q \otimes 1$ and Q_1 is the quotient space. Since Q is rationally injective, it is a direct summand of $Q \otimes k[G]$ by Lemma 1(b). Then $W \otimes Q$ is a direct summand of the rationally injective G-module $W \otimes (Q \otimes k[G])$. We now apply Lemma 1(a). QED

Theorem 3 [71; Corollary, p.179 or 45; Proposition 10.4, p.270]. *Let k be a field, X and Y irreducible algebraic k-schemes, and let $f: Y \to X$ be a morphism. Set $\dim X = n$, $\dim Y = m$, and suppose that the following conditions hold: (1) X is regular; (2) Y is Cohen-Macaulay; (3) f takes closed points of Y into closed points of X; (4) for every closed point $x \in X$ the fibre $f^{-1}(x)$ is $(m - n)$ dimensional. Then f is flat.*

Lemma 4 [71; Theorem 7.2(3), p.47]. *Let A be a ring and M an A-module. Then M is faithfully flat over A if and only if M is A-flat and $mM \neq M$ for every maximal ideal*

m of A.

Theorem 5. *Let H be a closed subgroup of G such that G/H is affine. Let* $A = k[G/H]$.
(a) $k[G] \otimes_A k[G] = k[G] \otimes k[H]$ *and the action of H on* $k[G] \otimes k[H]$ *given by right translation on k[G] and left translation on k[H] gives an action of H on* $k[G] \otimes_A k[G]$ *by right translation on the first k[G] term and the trivial action on the second.*
(b) *k[G] is a faithfully flat A-module.*

Proof. To prove statement (a), we shall use facts about the "fibred product" of schemes which may be found in [45, p.87]. First, since G/H is affine, $k[G] \otimes_A k[G]$ is the algebra of the fibred product $G \times_{G/H} G$ of G and G over G/H relative to the natural mappings $G \to G/H$. For $i = 1,2$, let p_i: $G \times_{G/H} G \to G$ be the projection maps. Next, let π_i: $G \times H \to G$ be given by $\pi_1(g,h) = g$ and $\pi_2(g,h) = gh$. We shall show that $G \times H$ is also the fibred product relative to the mappings π_i by checking the universal mapping property. To this end, let Z be a variety and let f_i: $Z \to G$ be morphisms so that $f_1(z)H = f_2(z)H$ for all $z \in Z$. We define a morphism θ: $Z \to G \times H$ by $\theta(z) = (f_1(z), f_1(z)^{-1}f_2(z))$. Then θ has the desired properties, namely, $f_1 = \pi_1 \circ \theta$, $f_2 = \pi_2 \circ \theta$ and θ is uniquely determined. Thus, $G \times H$ is the fibred product. It follows that $k[G] \otimes_A k[G]$ is isomorphic to $k[G] \otimes k[H]$. Let α: $k[G] \otimes_A k[G] \to k[G] \otimes k[H]$ be given by $\alpha(\Sigma f_i \otimes f_i')(g,h) = \Sigma f_i(g) f_i'(gh)$. Then α is the isomorphism since $\alpha \circ p_i^* = \pi_i^*$ (where * indicates the comorphism of a mapping). The fact about the H-actions now follows by a short calculation.

Now, we prove statement (b). To this end, let us recall part of the proof of Theorem 1.12, namely, there is a finite-dimensional G-module V_1 and a vector $z \in V_1$ so that $G^\circ \cdot z$ is isomorphic to $G^\circ/(H \cap G^\circ)$ and $G \cdot z$ is isomorphic to G/H. Since G/H is affine, so is $G^\circ/(H \cap G^\circ)$ as we see by considering the decomposition of G/H into irreducible components. Thus, the natural mapping from G° to $G^\circ/(H \cap G^\circ)$ is flat according to Theorem 3. It follows that the natural mapping from G to G/H is also flat. Then $k[G]$ is faithfully flat over $k[G/H]$ by Lemma 4 since G/H is affine. QED

Lemma 6. *Let* $A = k[G/H]$. *Let M be any A-module which is also a rational H-module such that* $h\cdot(am) = a(h\cdot m)$ *for all* $a \in A$, $h \in H$, *and* $m \in M$. *Let H act trivially on k[G]. Then* $(M \otimes_A k[G])^H = M^H \otimes_A k[G]$.

Proof. First, let L: $M \to M$ be any A-linear transformation. Then $\ker L \otimes_A k[G]$ is $\ker(L \otimes I)$, i.e., the kernel of the A-linear transformation $L \otimes I$ on $M \otimes_A k[G]$. Indeed, this follows immediately from tensoring with $k[G]$ the exact sequence of A-modules $0 \to \ker L \to M \to \mathrm{Image} L \to 0$. Now, if N_1 and N_2 are A-submodules of M, then $(N_1 \cap N_2) \otimes_A k[G] = (N_1 \otimes_A k[G]) \cap (N_2 \otimes_A k[G])$ according to [71; Theorem 7.4(i), p.48]. Thus, if L_i are finitely many A-linear transformations on M, then $\cap_i \ker(L_i \otimes I) = (\cap_i \ker L_i) \otimes_A I$.

Let K be any field which contains k. We put $M_K = M \otimes_k K$ and identify the k-rational points in M_K with $M = M \otimes 1$. In particular, let γ_i be generic points of the various components of H over k and let K be the field obtained by adjoining all the γ_i to k. Then, M^H consists of the k-rational points in $\cap M_K^{\gamma_i}$.

We apply what was just proved to $L_i = \gamma_i - I$ and use the result cited from [71] to see that $\cap_i (M_K \otimes_A k[G])^{\gamma_i} = \cap_i (M_K^{\gamma_i} \otimes_A k[G]) = (\cap_i M_K^{\gamma_i}) \otimes_A k[G]$. Now, $(M \otimes_A k[G])^H$ consists of all the k-rational points in $\cap_i (M_K \otimes_A k[G])^{\gamma_i} = ((\cap_i M_K^{\gamma_i}) \otimes_A k[G])$

$\cap (M \otimes_A k[G]) = (\cap_i(M_K^{\gamma_i} \cap M) \otimes_A k[G]) = M^H \otimes_A k[G]$. QED

Theorem 7. *Let H be a closed subgroup of G such that G/H is affine. Let $0 \to V_1 \to V_2 \to V_3 \to 0$ be an exact sequence of H-modules and H-homomorphisms. Then the sequence $0 \to ind_H{}^G V_1 \to ind_H{}^G V_2 \to ind_H{}^G V_3 \to 0$ is also exact.*

Proof. Let $A = k[G/H]$. We first shall show that if V is any H-module, then $(V \otimes k[G] \otimes k[H])^H \simeq ((V \otimes k[G])^H \otimes_A k[G])$ where H acts by left translation on $k[H]$, by right translation on the middle $k[G]$ terms, and trivially on the far right $k[G]$ term. Now, $V \otimes (k[G] \otimes k[H]) \simeq (V \otimes (k[G] \otimes_A k[G])$ by Theorem 5(a). By a standard fact about tensor products ([71; Formula 10, p.268]), this latter term is $(V \otimes k[G]) \otimes_A k[G]$. We now apply Lemma 6.

Let $0 \to V_1 \to V_2 \to V_3 \to 0$ be an exact sequence of H-modules and H-homomorphisms. We tensor this sequence over k with $k[G] \otimes k[H]$ to get another exact sequence. But $V_i \otimes k[G] \otimes k[H]$ is a rationally injective H-module by Theorem 2 and, by Lemma 1(b), the sequence $0 \to (V_1 \otimes k[G] \otimes k[H])^H \to (V_2 \otimes k[G] \otimes k[H])^H \to (V_3 \otimes k[G] \otimes k[H])^H \to 0$ is exact. But as we just saw, $(V_i \otimes k[G] \otimes k[H])^H \simeq (V_i \otimes k[G])^H \otimes_A k[G]$. Since $k[G]$ is faithfully flat over A (Theorem 5(b)), the sequence $0 \to (V_1 \otimes k[G])^H \to (V_2 \otimes k[G])^H \to (V_3 \otimes k[G])^H \to 0$ is exact. QED

We now prove the converse to Theorem 7 following an argument given in [29, p.247]. We begin with some lemmas.

Lemma 8. *A variety X is affine if and only if there is a finite set of elements $f_1, \dots, f_r \in k[X]$ such that the open sets $X_i = \{x \in X : f_i(x) \neq 0\}$ are affine and $f_1, \dots, f_r$ generate the unit ideal in $k[X]$.*

Proof. We need only prove that the existence of such f_i implies that X is affine. In doing this, we may assume that X is irreducible once we replace X by any one of its irreducible components. Then, $k[X_i] = k[X][1/f_i]$ (see Exercise 3). Now, choose elements $a_{ij} \in k[X]$ so that $k[X_i] = k[X][1/f_i] = k[a_{i1}, \dots, a_{ir(i)}][1/f_i]$ for all $i = 1, \dots, r$. Suppose, also, that (*) $1 = \Sigma_i p_i f_i$ where $p_i \in k[X]$. We shall show that $k[X] = k[a_{ij}, f_i, p_i]$. Indeed, let $f \in k[X]$. Then, for each i, we may find a non-negative integer $e(i)$ so that $ff_i^{e(i)} \in k[a_{ij}]$. We raise equation (*) to a high enough power that some f_i appears to at least the $e(i)$-th power in each monomial. Multiplying this new equation by f, we see that $f \in k[a_{ij}, f_i, p_i]$. Let Y be the affine variety so that $k[Y] = k[a_{ij}, f_i, p_i]$. Then, the open sets $Y_i = \{y \in Y : f_i(y) \neq 0\}$ and $X_i = \{x \in X : f_i(x) \neq 0\}$ are affine, have the same algebras of regular functions, and so are isomorphic. It follows that the map from X to Y corresponding to the inclusion of $k[Y]$ in $k[X]$ is an isomorphism. QED

It will be convenient to give here yet another equivalent condition for a closed subgroup H of G to be observable in G.

Definition. Let H be a closed subgroup of G. A rational H-module M is called *co-extendible* if and only if there is a short exact sequence of H-modules $0 \to Q \to V|_H \to M \to 0$ where V is a G-module.

Lemma 9. *Let H be a closed subgroup of G. Then H is observable in G if and only if*

for each one-dimensional, co-extendible H-module L, the dual H-module L^* is also co-extendible.

Proof. (Exercise 4).

Theorem 10. *If the induction functor is exact, then G/H is affine.*

Proof. First, we show that H is observable in G. Let L be a co-extendible H-module of dimension 1 with say, $0 \to Q \to V|_H \to L \to 0$ where V is a G-module. We tensor this exact sequence with L^* to obtain $0 \to Q \otimes L^* \to V|_H \otimes L^* \to L \otimes L^* \to 0$ where we note that $L \otimes L^*$ may be identified with the trivial H-module k. Since the induction functor is exact, we now have $0 \to \mathrm{ind}_H{}^G(Q \otimes L^*) \to \mathrm{ind}_H{}^G(V|_H \otimes L^*) \to \mathrm{ind}_H{}^G(k) \to 0$. The middle term in this exact sequence is isomorphic to $V \otimes \mathrm{ind}_H{}^G(L^*)$ according to (P5) and, then, $\mathrm{ind}_H{}^G(k) = k[G]^H$ by the Example, §6. In particular, $V \otimes \mathrm{ind}_H{}^G(L^*) \neq 0$ which, in turn, implies that $\mathrm{ind}_H{}^G(L^*) \neq 0$. The natural mapping E: $\mathrm{ind}_H{}^G(L^*) \to L^*$, defined just before the proof of (P1), is not zero (Exercise 1 below) and, so, is surjective since dim $L^* = 1$. Thus, if we let Z be the kernel of E, we have an exact sequence $0 \to Z \to \mathrm{ind}_H{}^G(L^*) \to L^* \to 0$. This shows that H is observable in G by Lemma 9. Thus, $X = G/H$ is quasi-affine (Theorem 2.1(4)). Let $X \subset Y$ where Y is an affine variety on which G acts morphically.

Let $f \in k[Y]$ be such that $X_f = \{x \in X : f(x) \neq 0\}$ is affine. Let $V = \langle G{\cdot}f \rangle$ be the finite-dimensional subspace of $k[Y]$ spanned by all the $g{\cdot}f$, $g \in G$. We note that $X_{g\cdot f} = g{\cdot}X_f$ is affine for all $g \in G$ since X_f is. Now, define a mapping φ: $k[X] = k[G]^H \to k$ by $\varphi(f) = f(eH)$ where, as usual e is the identity in G. We note that $\varphi(h{\cdot}f') = \varphi(f')$ for all $f' \in k[X]$ since $\varphi(h{\cdot}f') = (h{\cdot}f')(eH) = f'(h^{-1}eH) = f'(eH) = \varphi(f')$. Restricting φ to V, we obtain an exact sequence $0 \to K \to V \to k \to 0$ and, applying $\mathrm{ind}_H{}^G$, another exact sequence $0 \to \mathrm{ind}_H{}^G K \to \mathrm{ind}_H{}^G V \to k[G]^H \to 0$. We shall denote the mapping from $\mathrm{ind}_H{}^G V$ to $k[G]^H$ by φ^*; thus, for $\alpha \in \mathrm{ind}_H{}^G V$, we have $\varphi^*(\alpha)(g) = \alpha(g)(e)$.

Let $F \in k[G]^H$ be defined by $F(gH) = 1$ for all g and suppose that $F = \varphi^*(\alpha)$. Now, V is a G-module and we have seen (P7) that $\mathrm{ind}_H{}^G V$ is isomorphic to $k[G]^H \otimes V$ via a mapping ψ: $k[G]^H \otimes V \to \mathrm{ind}_H{}^G V$ given by $\psi(a \otimes b)(g) = a(g)g^{-1}b$. In particular, let $\alpha = \psi(\Sigma_i f_i \otimes f_i')$. We may assume that each f_i' has the form $g'{\cdot}f$ for suitable $g' \in G$. Then, applying φ^* to both sides of this equation, we see that for any $g \in G$ we have $\Sigma_i f_i(g)g^{-1}f_i'(e) = \Sigma_i f_i(g)f_i'(g) = \varphi^*(\alpha)(g) = F(g) = 1$. The conditions of Lemma 8 are satisfied by the f_i' so $X = G/H$ is affine. QED

Note. Using the notion of an "associated sheaf", one may approach induced modules in a different way [57; p.84]. Let H be a closed subgroup of G, let π: $G \to G/H$ be the canonical map and let V be an H-module. We define a sheaf $\mathcal{L}(V)$ on G/H as follows: if X is an open subset of G/H, put $\mathcal{L}(V)(X) = \{f \in \mathrm{Mor}(\pi^{-1}(X), V) : hf(gh) = f(g)$ for all $g \in \pi^{-1}(X)\}$. It may then be proved that (i) $\mathcal{L}$ is exact, (ii) $\mathcal{L}(V)$ is quasi-coherent and is coherent if V is finite-dimensional, (iii) $R^n\mathrm{ind}_H{}^G V \simeq H^n(G/H, \mathcal{L}_{G/H}(V))$. If G/H is affine, then $H^1(G/H, \mathfrak{I}) = 0$ for any coherent sheaf of ideals $\mathfrak{I}$ by Serre's theorem. It follows that induction is exact.

Exercises.

1. In the language of (P1), show that $E_V \neq 0$ if $\mathrm{ind}_H{}^G \neq 0$. (Hint: let I (resp. Z) be the identity (resp. zero) map on $\mathrm{ind}_H{}^G V$. If $E_V = 0$, then $E_V \circ I = E_V \circ Z$.)

2. Prove (P9).
3. Let X be irreducible and let $f \in k[X]$. Let $X_f = \{x \in X : f(x) \neq 0\}$. Show that $k[X_f] = k[X][1/f]$.
4. Prove Lemma 9. (Hint: apply Theorem 2.1(6) to the sequence of dual spaces.)
5. Let $H = TU$ be connected and solvable where T is a maximal torus in H and U is the unipotent radical of H. Let E be a T-module. Then $\mathrm{ind}_T^H E = k[U] \otimes E$ where U acts on $k[U]$ by left translation and T acts on $k[U]$ by $(t{\cdot}f)(u) = f(t^{-1}ut)$.

§ 7. Induced Modules and Observable Subgroups

In this section, we apply the results in §6 to the study of observable groups and homogeneous spaces.

Theorem 7.1. *Let L be an affine algebraic group and let H be a closed subgroup of L such that $\mathfrak{R}_u H \subset \mathfrak{R}_u L$. Then L/H is affine.*

Proof. Since $\mathfrak{R}_u L$ is a normal subgroup of L, the quotient space $L/\mathfrak{R}_u L$ is affine. Furthermore, $\mathfrak{R}_u L/\mathfrak{R}_u H$ is affine by Corollary 2.5. Thus, the induction functor from $\mathfrak{R}_u H$ to $\mathfrak{R}_u L$ and, also, from $\mathfrak{R}_u L$ to L is exact by (P10), §6. It follows (from (P4), §6) that the induction functor from $\mathfrak{R}_u H$ to L is exact; thus, $L/\mathfrak{R}_u H$ is affine according to (P10), §6. But the group $H/\mathfrak{R}_u H$ is reductive and, so (by Theorem A), the quotient space $L/H = (L/\mathfrak{R}_u H)/(H/\mathfrak{R}_u H)$ is affine. QED

Theorem 7.2 [92; Theorem A; p.38]. *Let H be a closed subgroup of a reductive group G. Then G/H is an affine variety if and only if H is a reductive group.*

Proof. If H is reductive, then G/H is affine (Corollary 4.6). To prove the converse, let $U = \mathfrak{R}_u H$ and suppose that $U \neq \{e\}$; we seek a contradiction. By assumption, G/H is affine. Since H/U is affine, we see that G/U is also affine (Exercise 2). Now, we may assume that G is a proper subgroup of some $SL(V)$. (Let ρ be a faithful representation of G on a vector space W, let W^* be the dual space of W and let G act on $V = W \oplus W^*$ by $(\rho(g), {}^t\rho(g^{-1}))$.) The homogeneous spaces $SL(V)/G$ and G/U are affine, so $SL(V)/U$ is also affine. We consider the action of U by left translation on the space $SL(V)/G$. The orbits of any unipotent affine group acting on any affine variety are all closed [10; Proposition 4.10, p.88]; it follows that all the double cosets GaU are closed in $SL(V)$. Then, the orbits of G acting on the affine variety $SL(V)/U$ are all closed (by Theorem A) since they are the images of the G-invariant closed sets GaU. Therefore, they have the same dimension. (Let $X = SL(V)/U$. The quotient mapping π: $X \to X/G$ separates orbits by Theorem A. Thus, π is an orbit map so the orbits of G have constant dimension [10; Proposition 6.4; p.95].) Let π: $SL(V) \to SL(V)/U$ be the quotient mapping and let $a \in SL(V)$. Then $G_{\pi(a)} = G \cap aUa^{-1}$ so $G_{\pi(e)} = U$. Thus, $\dim(G \cap aUa^{-1}) = \dim U$ for all $a \in SL(V)$. Since U is connected, $aUa^{-1} \subset G$ for all $a \in SL(V)$. The group generated by all the aua^{-1}, $u \in U$, is closed [10; Proposition 2.2; p.57]; it is obviously normal in $SL(V)$ and contained in G. But $SL(V)$ is almost simple so $G = SL(V)$. This contradiction completes the proof. QED

Let V be a finite-dimensional irreducible G°-module. Then $\mathfrak{R}_u G$ acts trivially on V. In fact, the non-zero subspace $V' = \{v \in V : g{\cdot}v = v \text{ for all } g \in \Re_u G\}$ is invariant under G° and, so, must be all of V. Thus, V is actually a $G^\circ/\Re_u G$-module. Since $G^\circ/\Re_u G$ is reductive, it makes sense to speak of highest weights in this context and we are able to make the following

Definition. A subgroup Q of G° is said to be *quasi-parabolic* if Q is the stabilizer of a highest weight vector in some finite-dimensional irreducible G°-module V. A subgroup H of G is called *subparabolic* if there is a quasi-parabolic subgroup Q of G° such that $H^\circ \subset Q$ and $\Re_u H \subset \Re_u Q$.

We note that any reductive subgroup of G is subparabolic since we can take the trivial representation of G in the definition. Our goal in the rest of this section is to prove the following group-theoretic characterization of observability due to Sukhanov [106].

Theorem 7.3. *A closed subgroup H of G is observable in G if and only if H is subparabolic.*

Proof. We may assume that both H and G are connected (Corollary 2.2). Let us first prove that the condition in the Theorem is sufficient. Suppose that H is subparabolic in Q, say. Then, H is observable in Q by Theorems 7.1 and 2.1(4). Since Q is observable in G (Theorem 2.1(2)), H is observable in G (Corollary 2.3). The proof of necessity requires some preparatory results.

Lemma 7.4 [10; Corollary 14.11, p.192]. *Let H and H' be linear algebraic groups and let $\pi: H \to H'$ be a surjective homomorphism. Then, $\pi(\Re_u H) = \Re_u(H')$.*

Lemma 7.5. *Let G act morphically on an affine variety X and let Y be a G-invariant closed subvariety of X. Then, there is a finite-dimensional rational G-module W and a G-equivariant morphism $f: X \to W$ such that $f^{-1}(0) = Y$.*

Proof. Let $I \subset k[X]$ be the ideal vanishing on Y. Let $f_1, \dots, f_r \in I$ be chosen so that (i) the f_i's are linearly independent, (ii) I is generated as an ideal by the f_i's and (iii) the space spanned by the f_i's is G-invariant. Suppose that $g{\cdot}f_j = \Sigma_i \rho(g)_{ij} f_i$. Let W be the dual space to the vector space spanned by the f_i's, i.e., W has a basis $w_1, \dots, w_r$ and $g{\cdot}w_j = \Sigma_i \rho(g)^{-1}{}_{ji} w_i$. We define $f: X \to W$ by $f(x) = \Sigma_i f_i(x) w_i$. Then f has the desired properties (Exercise 1). QED

The next result, due to F. A. Bogomolov, comes from geometric invariant theory; its proof is discussed in the appendix to this section.

Theorem 7.6 [9; Theorem 1, p.511]. *Let G be a connected reductive group and let V be a finite-dimensional G-module. Let $v \in V$, $v \neq 0$, be an element such that $0 \in cl(G{\cdot}v)$. Then, there is a proper quasi-parabolic subgroup Q of G so that $G_v \subset Q$.*

Lemma 7.7. *Let G be a (connected) reductive group and let H be a proper observable subgroup of G which is not reductive. There is a proper quasi-parabolic subgroup Q of G such that $H^\circ \subset Q$.*

Proof. According to Theorem 1.12, there is a finite-dimensional G-module V and a vector $v \in V$ such that $H = G_v$ and $G/H \simeq G{\cdot}v$. Let $X = \mathrm{cl}(G{\cdot}v)$ and let $Y = X - G{\cdot}v$. If Y is empty, then H is reductive by Theorem 7.2, a contradiction. Otherwise, let W and f: $X \to W$ be as in Lemma 7.5 and let $w = f(v)$. Then $H \subset G_w$ since f is G-equivariant. Since $0 \in \mathrm{cl}(G{\cdot}w)$, we may apply Theorem 7.6 to see that G_w is contained in a proper quasi-parabolic subgroup of G. QED

Lemma 7.8. *Let G be a reductive group and let H be an observable subgroup of G. Then H is subparabolic.*

Proof. As noted before, we may assume that both G and H are connected. If G is a torus or if H is reductive, the argument is immediate. We now proceed by induction on the dim G. According to Lemma 7.7, there is a proper quasi-parabolic subgroup Q of G such that $H \subset Q$. If the semi-simple rank of Q is equal to that of G, we may replace G by Q (Exercise 3) and apply the induction hypothesis. Otherwise, $\mathfrak{R}_u Q \neq \{e\}$. For if $\mathfrak{R}_u Q = \{e\}$, then Q is reductive and since Q contains U, it has the same semi-simple rank as G. Next, we show that $\mathfrak{R}_u H \subset \mathfrak{R}_u Q$. In doing this, we may assume that $H \supset \mathfrak{R}_u Q$. Indeed, the group $H{\cdot}\mathfrak{R}_u Q$ is observable in Q according to Corollary 2.11. Since Q is observable in G, we see that $H{\cdot}\mathfrak{R}_u Q$ is observable in G (Corollary 2.3) and that $\mathfrak{R}_u H \subset \mathfrak{R}_u(H{\cdot}\mathfrak{R}_u Q)$ (Exercise 4). Thus, from now on, we may assume that $H \supset \mathfrak{R}_u Q$.

Since H is observable in Q, the variety Q/H is quasi-affine (Theorem 2.1(4)). Thus, $H/\mathfrak{R}_u Q$ is observable in $Q/\mathfrak{R}_u Q$ since $Q/H \simeq (Q/\mathfrak{R}_u Q)/(H/\mathfrak{R}_u Q)$. By induction, there is a quasi-parabolic subgroup Q' of $Q/\mathfrak{R}_u Q$ such that $H/\mathfrak{R}_u Q$ is subparabolic in Q'. Consider the natural mapping π: $Q \to Q/\mathfrak{R}_u Q$ and let $L = \pi^{-1}(Q')$. Then, L is observable in Q (by Theorem 2.1(2)) since $\pi(L) = Q'$ is observable in $\pi(Q)$. Since Q is observable in G, L is also observable in G. Furthermore, since $\pi(L) = Q'$ is normalized by a maximal torus of $Q/\mathfrak{R}_u Q$, L is normalized by a maximal torus in G. Indeed, there is a parabolic subgroup P of G such that $\mathfrak{R}_u Q = \mathfrak{R}_u P$, Q is normal in P and dim $(P/Q) = 1$; the subgroup $Q/\mathfrak{R}_u Q$ has the same semi-simple rank as $P/\mathfrak{R}_u Q$. By Theorem 3.9, L is subparabolic. But $\mathfrak{R}_u H \subset \mathfrak{R}_u L$; for by Lemma 7.4, we have $\pi(\mathfrak{R}_u H) = \mathfrak{R}_u(\pi(H)) \subset \mathfrak{R}_u Q' = \mathfrak{R}_u(\pi(L)) = \pi(\mathfrak{R}_u L)$. Hence, H is subparabolic. QED

We now prove that the condition stated in Theorem 7.3 is necessary. Again, we may assume that both G and H are connected. We may also assume that $H \supset \mathfrak{R}_u G$. Indeed, the group $H{\cdot}\mathfrak{R}_u G$ is observable in G by Corollary 2.11. Furthermore, $\mathfrak{R}_u H \subset \mathfrak{R}(H{\cdot}\mathfrak{R}_u G)$ (Exercise 4). Then, $H_1 = H/\mathfrak{R}_u G$ is observable in $G_1 = G/\mathfrak{R}_u G$ by Theorem 2.1(4) since $G_1/H_1 \simeq G/H$. According to Lemma 7.8, H_1 is subparabolic in G_1. Let π: $G \to G_1$ be the canonical map. Let ρ: $G_1 \to GL(V)$ be an irreducible finite-dimensional representation and let v be a non-zero highest weight vector in V so that $H_1 \subset (G_1)_v = Q_1$ and $\mathfrak{R}_u H \subset \mathfrak{R}_u Q_1$. Then $\rho \circ \pi$ is an irreducible representation of G so that $H \subset Q = G_v$. Furthermore, $\mathfrak{R}_u H \subset \mathfrak{R}_u Q$ since $\pi(\mathfrak{R}_u H) = \mathfrak{R}_u \pi(H) \subset \mathfrak{R}_u \pi(Q) = \pi(\mathfrak{R}_u Q)$. QED

Appendix.

Our purpose here is to sketch a proof of Theorem 7.6. In so doing, we shall generally follow the discussion given in [91].

(1) [10; §8,p.111]. Let G be a reductive group, let T be a maximal torus in G and let

$X(T)$ be the character group of T. We choose an inner product on $X(T)$, say (χ, χ'), which is (i) invariant under the Weyl group of T, (ii) is **Q**-valued, and (iii) makes the central torus and the semi-simple torus of G orthogonal. This inner product determines an inner product for any other maximal torus since any two maximal tori are conjugate.

Let $X_*(T)$ (resp. $X_*(G)$) be the group of all one-parameter subgroups of T (resp. G), i.e., homomorphisms from k^* to T (resp. G). There is a mapping from $X(T) \times X_*(T) \to \mathbf{Z}$, $(\chi, \lambda) \to <\chi, \lambda>$, which is a dual pairing over **Z**. The inner product on $X(T)$ gives an inner product on $X_*(T)$, say (λ, λ'). For $\lambda \in X_*(T)$, we define $\chi_\lambda \in X(T)$ by $(\lambda, \lambda') = \chi_\lambda(\lambda')$ for all $\lambda' \in X_*(T)$. Then, $(\lambda, \lambda') = \chi_\lambda(\lambda') = (\chi_\lambda, \chi_{\lambda'})$. Since any two maximal tori in G are conjugate, for any one-parameter subgroup λ of G, we have a well defined norm $\|\lambda\|$ with $\|\lambda\|^2 \in \mathbf{Q}$.

(2) [74; Chapter 2, §2, p.55] For $\lambda \in X_*(T)$, we define a corresponding parabolic subgroup $P(\lambda)$ by $P(\lambda) = \{g \in G : \text{limit}_{a \to 0} \lambda(a)g\lambda(a)^{-1} \text{ exists in } G\}$. The subgroup $P(\lambda)$ has a Levi-decomposition $P(\lambda) = L(\lambda)U(\lambda)$ where $L(\lambda) = \{g \in G : \lambda(a)g\lambda(a)^{-1} = g\}$ is the reductive group and $U(\lambda) = \{g \in G : \text{limit}_{a \to 0} \lambda(a)g\lambda(a)^{-1} = e\}$ is the unipotent radical. The group $L(\lambda)$ (resp. $U(\lambda)$) is normalized by T and is generated by those roots $\alpha \in \Phi(T,G)$ such that $<\alpha, \lambda> = 0$ (resp. $<\alpha, \lambda> > 0$).

(3) [91; Section 1, p.270] Let V be a finite-dimensional G-module and let $\chi \in X(T)$. We define $V_\chi = \{v \in V : t{\cdot}v = \chi(t)v \text{ for all } t \in T\}$; then, $V = \oplus V_\chi$ where the sum is over all those characters χ such that $V_\chi \neq \{0\}$. For $v \in V$, we have a corresponding decomposition $v = \oplus v_\chi$. We define the *state of* v, denoted by $S_T(v)$, to be $\{\chi : v_\chi \neq 0\}$.

For $\lambda \in X_*(T)$, we have $V = \oplus_{i \in \mathbf{Z}} V_i$ where $V_i = \{v \in V : \lambda(a)v = a^i v \text{ for all } a \in k^*$. Thus, $V_i = \oplus V_\chi$ where the sum is over all those characters χ such that $<\chi, \lambda> = i$. Let $V^q = \oplus_{i \geq q} V_i$. Then, $V^{q+1} \subset V^q$ and each V^q is invariant under the action of the group $P(\lambda)$ by Lemma 3.1. Thus, the group $P(\lambda)$ acts on the quotient space V^{q+1}/V^q and the canonical mapping π from V^{q+1} to V^{q+1}/V^q is $P(\lambda)$-equivariant. We also note that $U(\lambda)$ acts trivially on the quotient V^{q+1}/V^q by Lemma 3.1, again.

(4) From now on, we fix an element $v_o \in V$ such that $0 \in \text{cl}(G{\cdot}v_o)$.

Theorem [74; Theorem 2.1, p.49]. *There is a one-parameter subgroup λ of G such that $\text{limit}_{a \to 0} \lambda(a){\cdot}v_o = 0$.*

Actually, we shall focus on a special family of one-parameter subgroups of G, found by G. Kempf, which satisfy the condition in the theorem. For a one-parameter subgroup λ of G whose image is contained in some maximal torus T', we define $\mu(v_o, \lambda) = \inf\{(<\chi, \lambda> : \chi \in S_{T'}(v_o)\}$. Kempf proved that the function $\lambda \to \mu(v_o, \lambda)/\|\lambda\|$ attains a maximum value on $X_*(G)$. After renaming, we may suppose that such a λ is in T. Then, λ is uniquely determined among the indivisible one-parameter subgroups of T and is called the *instability one-parameter subgroup for* v_o [91; p.274]. We put $j = \mu(v_o, \lambda)$; then $j = \max\{q : v_o \in V^q\}$.

Theorem [91; Proposition 1.12, p.276]. *Let λ be the instability one-parameter subgroup for v_o. There is non-negative integer d and a non-constant homogeneous function f on*

V^j/V^{j+1} *such that* $f(\pi(v_o)) \neq 0$ *and* $f(p\cdot\pi(v)) = \chi_\lambda(p)^d f(\pi(v))$ *for all* $v \in V^j$ *and* $p \in P(\lambda)$.

5) According to Corollary 3.6, there is a finite-dimensional, irreducible representation W of G having highest weight vector w with T-weight $d\chi_\lambda$ such that $G_w = P(\lambda)$. Now, consider the mapping φ from $G \times V^j$ to W given by $\varphi(g, v) = f(\pi(v))g\cdot w$. The group G acts on $G \times V^j$ by $x\cdot(g, v) = (xg, v)$ and φ is G-equivariant. The group $P(\lambda)$ acts on $G \times V^j$ by $p\cdot(g, v) = (gp^{-1}, p\cdot v)$ and φ is $P(\lambda)$-invariant since $\varphi(p\cdot(g, v)) = \varphi(gp^{-1}, p\cdot v) = f(\pi(p\cdot v))gp^{-1}\cdot w = f(p\cdot\pi(v))gp^{-1}\cdot w = \chi_\lambda(p)^d f(\pi(v))g\chi_\lambda(p)^{-d}w = \varphi(g, v)$. Also, $\varphi(e, v_o) = f(\pi(v_o))e\cdot w = cw$ for some $c \in k^*$; modifying w by a non-zero constant, we may assume that $\varphi(e, v_o) = w$.

Now, there is a mapping from $G \times V^j$ to V, given by $(g, v) \to g\cdot v$, which is G-equivariant. Its image, say Z, contains v_o and is closed by Theorem 5.3 since $Z = G\cdot V^j$ and $P(\lambda)\cdot V^j \subset V^j$.

Theorem [9; p.513]. *The map* φ *factors through Z, say,* $\varphi(g,v) = \psi(g\cdot v)$ *where* $\psi\colon Z \to W$ *is a morphism.*

Theorem 7.6 [9; pp.513-514]. *Let G be a connected, reductive group and let V be a finite-dimensional G-module. Let* $v \in V$, $v \neq 0$, *be an element such that* $0 \in cl(G\cdot v)$. *Then, there is a proper quasi-parabolic subgroup Q of G so that* $G_v \subset Q$.

Proof. We construct the maps φ and ψ as above (with v now playing the role of v_o). Then $G_v \subset G_w$ since if $g \in G_v$, we have $g\cdot w = g\cdot\varphi(e, v) = \varphi(g, v) = \psi(g\cdot v) = \psi(v) = \varphi(e, v) = w$. QED

Exercises.

1. Show that the mapping f defined in the proof of Lemma 7.5 is G-equivariant.

2. Let H_1 and H_2 be closed subgroups of G such that $H_1 \subset H_2$. Suppose that both H_2/H_1 and G/H_2 are affine. Show that G/H_1 is affine.

3. Prove the statement in the first paragraph of Lemma 7.8. To do this, one must prove the following. Let G and G_1 be reductive groups such that G is a subgroup of G_1 and suppose that G and G_1 have the same semi-simple rank. Then any finite-dimensional, irreducible representation of G extends to a finite-dimensional, irreducible representation of G_1. (Hint: let G act irreducibly on the finite-dimensional vector space V. Since G is observable in G_1, we may suppose that V is contained in a finite-dimensional G_1-module W. Let T be the center of G_1. Let $W = \oplus W_\chi$ be the decomposition of W into T-weight spaces. Then V is isomorphic to a subspace of one of the W_χ's.)

4. Let H be a closed subgroup of G and let $L = H\cdot\mathfrak{R}_u G$. Show that $\mathfrak{R}_u H \subset \mathfrak{R}_u(L)$. (Hint: consider the natural mapping $L \to L/\mathfrak{R}_u L$. The image of H is the reductive group $L/\mathfrak{R}_u L$. Now, apply Lemma 7.4.)

§8. Counter-examples

Induced modules provide a mechanism for generating counter-examples to Hilbert's fourteenth problem once examples are known for $\mathbf{G}_a$. Thus, we begin with three examples of $\mathbf{G}_a$ actions where the algebra of invariants is not finitely generated over k.

An elementary example [37; Example 2.3, p.82]. Let $A = k[\epsilon,x]$ where $\epsilon^2 = 0$. Let $\mathbf{G}_a$ act on A by $t{\cdot}f = f + t\epsilon x(\partial f/\partial x)$. Then one checks easily that this is a rational action of $\mathbf{G}_a$ and that the algebra of $\mathbf{G}_a$-invariants is $k[\epsilon x, \epsilon x^2, \ldots]$ which is not finitely generated over k.

The Nagata counter-example [104 and, also, 25,27,75]. We begin with some notation which will be used later when we refer to this example. Let $G = SL(2,k)$. Let B (resp. B^-) be the subgroup of G consisting of upper (resp. lower) triangular matrices. Let T (resp. U) be the subgroup of B consisting of diagonal matrices (resp. matrices with 1's on the diagonal). Let G act on $\mathbf{A}^2$, the space of 2×1 column matrices over k, in the usual way. Let $G_1 = G \times \ldots \times G$ (9 copies); the action of G on $\mathbf{A}^2$ gives an action of G_1 on $V = \mathbf{A}^2 \times \ldots \times \mathbf{A}^2$ (9 copies). We define subgroups of G_1 as follows: $T_1 = T \times \ldots \times T$, $U_1 = U \times \ldots \times U$, $B_1 = B \times \ldots \times B$, and $B_1^- = B^- \times \ldots \times B^-$. Let D be the subgroup of G_1 which consists of all elements of the form

$$\begin{pmatrix} c & 0 \\ 0 & 1/c \end{pmatrix} \times \cdots \times \begin{pmatrix} c & 0 \\ 0 & 1/c \end{pmatrix}.$$

Let $a_1, \ldots a_9$ be distinct non-zero elements in k such that $\Sigma_i\, a_i \neq 0$ and the product $a_1 \ldots a_9$ is not an nth root of unity for any n. Let H be the subgroup of G_1 which consists of the product of 9 matrices each of which has the form

$$\begin{pmatrix} 1 & b_i \\ 0 & 1 \end{pmatrix}$$

where the b_i satisfy the following three equations: (1) $\Sigma_i\, b_i = 0$, (2) $\Sigma_i\, a_i b_i = 0$ and (3) $\Sigma_i\, (a_i^2 - a_i^{-1})b_i = 0$. We note that the subgroup D normalizes H. Robert Steinberg greatly simplified and generalized Nagata's original counter-example and showed that the algebra of invariants for H acting on V is not finitely generated.

Next, let $H_o = \{e\} \subset H_1 \subset \ldots \subset H_n = H$ be a sequence of subgroups in H so that H_i is normal in H_{i+1} and $H_{i+1}/H_i \simeq \mathbf{G}_a$. Let V be as above. For each $i = 0,1,\ldots,n$, let A_i be the algebra of H_i-invariants on $k[V]$. Since H_i is normal in H_{i+1}, the group $\mathbf{G}_a \simeq H_{i+1}/H_i$ acts on A_i and A_{i+1} is the algebra of its invariants. We note that $A_o = k[V]$ is finitely generated over k, but $A_n = k[V]^H$ is not. Let m be the first index so that A_m is finitely generated over k but A_{m+1} is not. Then the algebra of invariants of $\mathbf{G}_a$ acting on A_m is not finitely generated over k.

The counter-examples of P. Roberts [93]. The counter-examples found by P. Roberts are for actions of $\mathbf{G}_a$ on $\mathbf{C}^7$ defined in the following manner. Fix an integer $m \geq 2$ and define an action by $r{\cdot}(x, y, z, s, t, u, v) = (x, y, z, s + rx^{m+1}, t + ry^{m+1}, u + rz^{m+1},$

$v + r(xyz)^m$). Paul Roberts showed that the algebra of invariants, S, for this action is not finitely generated over k. Annette A'Campo-Neuen extended this example and showed that there is a linear action of $(\mathbf{G}_a)^{12}$ on $\mathbf{A}^{19}$ whose algebra of invariants is the polynomial algebra in one variable over S and, so, is not finitely generated [1].

We next show that these counter-examples extend to give counter-examples for any non-reductive group.

Lemma 8.1. *Let G be an affine algebraic group and let H be a closed subgroup of G such that G/H is affine. Let A be a rational H-algebra and let $R = \mathrm{ind}_H{}^G A$. Then the algebras R^G and A^H are isomorphic. Furthermore, if A is finitely generated over k, then so is R.*

Proof. The group G (resp. H) acts on $k[G] \otimes A$ by left translation on $k[G]$ and the trivial action on A (resp. right translation on $k[G]$ and the given action on A). The actions of G and H commute so $R^G = ((k[G] \otimes A)^H)^G = ((k[G] \otimes A)^G)^H$. To prove the first statement, it suffices to show that $(k[G] \otimes A)^G$ and A are isomorphic as H-modules. Now consider the mapping φ: $k[G] \otimes A \to A$ given by $\varphi(\Sigma_i f_i \otimes a_i) = \Sigma_i f_i(e)a_i$. We shall show that φ is an H-equivariant isomorphism. The mapping is surjective since for each $a \in A$, we have $1 \otimes a \to a$. To show that φ is injective, suppose that $\varphi(\Sigma_i f_i \otimes a_i) = 0$ where we may assume that the a_i's are linearly independent over k. Then each of the $f_i(e) = 0$. But $\Sigma_i f_i \otimes a_i$ is G-invariant so that $\Sigma_i \ell_g f_i \otimes a_i = \Sigma_i f_i \otimes a_i$ for all $g \in G$. Hence, each $f_i(g) = 0$ and $f_i = 0$. Finally, let $h \in H$ and let $\Sigma_i f_i \otimes a_i \in (k[G] \otimes A)^G$. Then, $h\cdot(\Sigma_i f_i \otimes a_i) = h\cdot\Sigma_i (\ell_h f_i \otimes a_i) = \Sigma_i r_h \ell_h f_i \otimes h\cdot a_i \to \Sigma_i f_i(e) h\cdot a_i = h\cdot(\Sigma_i f_i(e)a_i)$ so that φ is H-equivariant.

To prove the second statement, we show first that there is a finite-dimensional H-module W and an H-equivariant surjective homomorphism π from $k[W]$ to A. Indeed, suppose that $A = k[a_1,\ldots,a_m]$ where the a_i's are linearly independent and that $h\cdot a_j = \Sigma_i t_{ij}(h)a_i$ for suitable $t_{ij} \in k[H]$. Let $x_1,\ldots,x_m$ be indeterminates over k, define $h\cdot x_j = \Sigma_i t_{ij}(h)x_i$, and let W^* have basis $\{x_1,\ldots,x_m\}$. Then, the mapping π: $k[W] \to A$ given by $\pi(x_i) = a_i$ satisfies the desired properties. Next, applying Theorem 2.1(7), we may assume that $W \subset V$ where V is a finite-dimensional G-module. The mapping from $k[V]$ to A (given by restriction to $k[W]$ and then the homomorphism above) is H-equivariant and surjective; let I be its kernel. The exact sequence $0 \to I \to k[V] \to A \to 0$ gives rise to an exact sequence $0 \to \mathrm{ind}_H{}^G I \to \mathrm{ind}_H{}^G k[V] \to \mathrm{ind}_H{}^G A \to 0$ according to (P10), §6. Since $k[V]$ is a G-module, $\mathrm{ind}_H{}^G k[V] = k[G/H] \otimes k[V]$ by (P6), §6, and, so, is finitely generated over k. Hence, $R = \mathrm{ind}_H{}^G A$ is also finitely generated over k. QED

The next result is essentially due to V. L. Popov [87] except we have translated his argument into the setting of induced representations.

Theorem 8.2. *Let G be an affine algebraic group. Then the following statements are equivalent; (i) G is reductive, (ii) for each finitely generated, commutative, rational G-algebra A, the algebra of invariants A^G is finitely generated over k.*

Proof. That statement (i) implies (ii) is the content of Theorem A. To prove the converse, let us assume that G is not reductive. Then, there is a connected unipotent abelian normal subgroup L of G. Let $H \simeq \mathbf{G}_a$ be a subgroup of L. Let A be any finitely generated, commutative k-algebra on which H acts in such a way that A^H is not finitely generated. Let $S = \mathrm{ind}_H{}^L A$ and let $R = \mathrm{ind}_L{}^G S = \mathrm{ind}_H{}^G A$. By Lemma 8.1, the algebra S is finitely generated over k but $S^L = A^H$ is not. Similarly, the algebra R is finitely generated over k but $R^G = S^L = A^H$ is not. QED

In the remainder of this section, we sketch some facts relevant to $\mathbf{G}_a$-actions and, especially, the counter-examples discovered by P. Roberts.

- We shall always assume that char k = 0.

Definition. Let A be a commutative k-algebra. *A locally nilpotent derivation* δ of A is a k-linear mapping from A to A such that (i) $\delta(ab) = a\delta(b) + b\delta(a)$ for all $a,b \in A$ and (ii) for each $a \in A$, there is a positive integer n so that $\delta^n a = 0$.

We show now that $\mathbf{G}_a$-actions correspond to locally nilpotent derivations. Indeed, let $\mathbf{G}_a$ act rationally on a commutative k-algebra A. Let $f \in A$. Then f belongs to a finite-dimensional vector space V which is invariant under $\mathbf{G}_a$ so there are elements $f_i \in A$, $i = 1, \dots, m$ so that $a{\cdot}f = f + af_1 + \dots + a^m f_m$ for all $a \in \mathbf{G}_a$. We define $\delta(f)$ to be f_1. If $g \in A$ and if $a{\cdot}g = g + ag_1 + \dots + a^p g_p$, then $a{\cdot}(fg) = (a{\cdot}f)(a{\cdot}g) = fg + a(fg_1 + gf_1) + \dots$ so that $\delta(fg) = fg_1 + gf_1 = f\delta(g) + g\delta(f)$ so δ is a derivation of A. Next, for all $a,b \in k$, we have $(a + b){\cdot}f = f + (a + b)f_1 + \dots + (a + b)^m f_m$; but $(a + b){\cdot}f = a{\cdot}(b{\cdot}f) = a{\cdot}(f + bf_1 + \dots + b^m f_m) = a{\cdot}f + b(a{\cdot}f_1) + \dots + b^m(a{\cdot}f_m)$. Comparing coefficients of ab, we see that $2f_2 = \delta(f_1)$, i.e., $f_2 = (1/2)\delta(f_1) = (1/2)\delta^2(f)$. In general, comparing coefficients of ab^{r-1}, we see that $f_r = (1/r!)\delta^r(f)$. Therefore, (*) $a{\cdot}f = f + a\delta(f) + (a^2/2!)\delta^2(f) + \dots + (a^m/m!)\delta^m(f)$ so δ is locally nilpotent. On the other hand, if δ is a locally nilpotent derivation of A, we define $a{\cdot}f$ by equation (*) and check that this gives an action of $\mathbf{G}_a$. The following lemma is an immediate consequence of the definitions.

Lemma 8.3. *Let A be a rational G_a-algebra and let δ be the corresponding locally nilpotent derivation. Let $f \in A$. Then $a{\cdot}f = f$ for all $a \in G_a$ if and only if $\delta(f) = 0$.*

Now, we sketch the argument behind the counter-examples discovered by Roberts. Let $R = \mathbf{C}[x,y,z,s,t,u,v]$ be a polynomial algebra in 7 indeterminates. We define a degree function on R by setting the degree of x, y, and z to be 0 and of s, t, u, v to be 1. Let R_n be the elements in R having degree n. Next, let $m \geq 2$ be an integer. To define a derivation δ of R, it suffices to give its action on the indeterminates. We put $\delta(x) = \delta(y) = \delta(z) = 0$, $\delta(s) = x^{m+1}$, $\delta(t) = y^{m+1}$, $\delta(u) = z^{m+1}$, and $\delta(v) = (xyz)^{m+1}$. We note that if $f = \Sigma_n f_n$ with $f_n \in R_n$, then $\delta(f) = 0$ if and only if each $\delta(f_n) = 0$. Also, δ is locally nilpotent and, so, we have a $\mathbf{G}_a$ action.

We consider R as a polynomial algebra in the variables s, t, u, v over the algebra $R_0 = k[x, y, z]$. Let $M = (x, y, z)$ be the ideal in R_0 generated by x, y, and z.

Lemma 8.4. *Let $f \in R_n$ be such that $\delta(f) = 0$. Then, the coefficients of f are in the ideal M.*

Proof. Suppose that this is not true. We may assume that $f = (1 + m)s^a t^b u^c v^d + \dots$

with, say, $a > 0$ and $m \in M$. Then, $\delta(f) = [(1 + m)ax^{m+1} + r]s^{a-1}t^b u^c v^d + \ldots$ where r is in the ideal in R_0 generated by y^m and z^m. But, then, $\delta(f) \neq 0$ since x^{m+1} is not of the form $mx^{m+1} + r$. QED

Lemma 8.5. *For each $n \geq 1$, there is an invariant $f \in R_n$ of the form $f = xv^n + \ldots$.*

Lemma 3 is, in fact, the key result in Roberts' paper. It can be proved by arguments from linear equations as in [93; pp 466-472] or from the standpoint of differential equations [24; p.6300]. In case $n = 2$, the first few invariants described in Lemma 8.5 are as follows [24; p.6299]:

$$
\begin{aligned}
c_1 &= vx - sy^2z^2, \\
c_2 &= v^2x - 2vsy^2z^2 + stx^2yz^4 + sux^2y^4z - tux^5yz, \\
c_3 &= v^3x - 3v^2sy^2z^2 + 3vstx^2yz^4 + 3vsux^2y^4z - \\
&\quad 3vtux^5yz - st^2x^4z^6 - stux^4y^3z^3 - su^2x^4y^6 + \\
&\quad t^2ux^7z^3 + tu^2x^7y^3.
\end{aligned}
$$

Let us assume that Lemma 8.5 holds and finish the argument.

Theorem 8.6. *The algebra of $R^\delta = \{r \in R : \delta(f) = 0\}$ is not finitely generated over k.*
Proof. Suppose that R^δ is finitely generated, say $R^\delta = k[f_1, \ldots, f_r]$. We may assume that f_i is homogeneous of degree $d(i)$. Then, by Lemma 8.4, we have $f_i = r_0 v^{d(i)} + \ldots$ as a polynomial in s, t, u, v over R_0 where the coefficients are in the ideal (x, y, z). Let $n > \max\{d(i) : i = 1, \ldots, r\}$ and let $f = xv^n + \ldots$ be as in Lemma 3. If f were a polynomial in the f_i's, then the coefficient of v^n would be in M^2. But x is not in M^2. QED

§9. The Transfer Principle

The transfer principle was proved in §6 as part of a sequence of properties of induced modules (namely, (P8)). Since this result is so important to what follows, we shall give an elementary, self-contained proof here.

Theorem 9.1. *Let Z be a rational G-module and let $(k[G] \otimes Z)^G = \{\Sigma f_i \otimes z_i \in k[G] \otimes Z : \Sigma \ell_g f_i \otimes g{\cdot}z_i = \Sigma f_i \otimes z_i$ for all $g \in G\}$ where, as usual, ℓ_g denotes the action of G on k[G] by left translation. Let Φ: $(k[G] \otimes Z)^G \to Z$ be defined by $\Phi(\Sigma f_i \otimes z_i) = \Sigma f_i(e)z_i$.*
(a) *The mapping Φ is a vector space isomorphism. If Z is a k-algebra, Φ is an algebra isomorphism.*
(b) *Let H be a subgroup of G and let H act on k[G] (and, hence, on $(k[G] \otimes Z)^G$) by right translation, denoted as usual by r_h. Then Φ is H-equivariant. In particular, $(k[G]^H \otimes Z)^G \cong Z^H$.*
Proof. To prove (a) we first show that Φ is injective. Let $b = \Sigma f_i \otimes z_i \in (k[G] \otimes Z)^G$ be in the kernel of Φ; we may assume that the z_i's are linearly independent. Let $g \in G$. Since $g{\cdot}b = b$, we have $\Phi(g{\cdot}b) = \Sigma f_i(g^{-1})g{\cdot}z_i = 0$. But, the $g{\cdot}z_i$'s are linearly

independent, so, each $f_i(g^{-1}) = 0$. Since g may be chosen arbitrarily in G, each $f_i = 0$. Next, we show that Φ is surjective. Let $z \in Z$. Since the action of G on Z is locally finite, there are elements $z_1,\ldots,z_m$ in Z and functions $t_{ij} \in k[G]$ such that (i) the space spanned by the z_i's is G-invariant and contains z, (ii)

$$g \cdot z_j = \Sigma_{i=1}^{m} t_{ij}(g) z_i$$

for each $j = 1,\ldots,m$ and all $g \in G$. We note that $\ell_g \cdot t_{ij} = \Sigma t_{ir}(g^{-1})t_{rj}$. Suppose that $z = \Sigma c_i z_i$ and let $f_i = \Sigma t_{ij} c_j$. Then, a straightforward computation shows that $\Sigma f_i \otimes z_i$ is G-invariant and is sent by Φ to z.

Since left and right translation commute, H sends $(k[G] \otimes Z)^G$ to itself. Thus, to prove (b), let $b = \Sigma f_i \otimes z_i \in (k[G] \otimes Z)^G$ and let $h \in H$. Then, $\Phi(\Sigma r_h \cdot f_i \otimes z_i) = \Sigma f_i(h) z_i$. Since b is G-invariant, $\Phi(b) = \Sigma f_i(e) z_i = \Phi(h^{-1} \cdot b) = \Sigma f_i(h) h^{-1} \cdot z_i$. It follows that Φ is H-equivariant. QED

Note. In the introduction to this chapter, we gave some of the history behind Theorem 9.1. To continue that, we note that the formulation above is taken from [6; Proposition 1, p.1341]. The rediscovery of the "adjunction argument" in recent times seems to have started with [38; p.245], but also see [113; Proposition 1, p.7], [85; p.10] and [89; Theorem 4, p.319].

Ordinarily, we shall not be interested in rationality questions. However, in the context of some of the examples in §11, it will be nice to have such an interpretation of Theorem 9.1. We begin by recalling some of the relevant terminology and facts from [10]. Let K be a subfield of k and V be a K-variety. We denote by $V(K)$ the set of K-rational points of V and by $K[V]$ the K-algebra of regular functions defined over K on V. If G is a K-group and if H is a closed subgroup of G defined over K, then the quotient G/H is defined over K. Let Z be a k-vector space which has a K-structure defined by Z_K which is a K-subspace. If G is a K-group which acts rationally on Z, we say that this action is defined over K if every element of Z_K is contained in a finite-dimensional subspace of Z which is defined over K, is invariant under G, and on which G acts by a rational representation which is defined over K. We recall that if K is an infinite, perfect field and if G is connected, then $G(K)$ is Zariski-dense in G [10; 18.3].

Theorem 9.2. *Let K be an infinite, perfect subfield of k. In the context of Theorem 9.1, suppose that G, H, Z, and the action of G on Z are all defined over K. Then Φ is also defined over K and gives an isomorphism from $(K[G/H] \otimes_K Z_K)^{G(K)}$ to $Z_K^{H(K)}$.*

Proof. That Φ is defined over K is immediate. If $a \in (K[G/H] \otimes_K Z_K)^{G(K)}$, then $g \cdot a = a$ for all $g \in G$ since $G(K)$ is Zariski-dense in G. Then $\Phi(a) \neq 0$ by Theorem 9.1. To see that Φ is surjective, let $z \in Z_K^{H(K)}$. Since $H(K)$ is Zariski-dense in H, we see that $z \in Z^H$. The element b, constructed in the proof of Theorem 9.1, with $\Phi(b) = z$ is K-rational. QED

We now begin in earnest our study of the application of the transfer principle to questions involving finite generation.

Theorem 9.3. *Let G be a reductive algebraic group and let H be a closed subgroup of G. The following conditions are equivalent:*
(a) *$k[G]^H$ is finitely generated over k;*
(b) *if A is any finitely generated, commutative k-algebra on which G acts rationally, then A^H is finitely generated over k.*
In case H is observable, conditions (a) and (b) are equivalent to:
(c) *there is a finite-dimensional, rational representation $\rho: G \to GL(V)$ and a vector $v \in V$ such that $H = G_v$ and each irreducible component of $cl(G{\cdot}v) - G{\cdot}v$ has codimension ≥ 2 in $cl(G{\cdot}v)$, i.e., H is a Grosshans subgroup of G.*

Proof. Statement (a) follows immediately from (b) when we let G act on $k[G]$ by right translation. If (a) holds, we may apply Theorem 9.1 to see that $A^H = (k[G/H] \otimes A)^G$. By assumption, $k[G/H]$ is finitely generated. Hence, $(k[G/H] \otimes A)^G$ is finitely generated over k according to Theorem A. The equivalence of statements (a) and (c) in the case where H is observable is the content of Theorem 4.3. QED

Note. Not all subgroups of reductive groups G satisfy the conditions in Theorem 9.3. In fact, let H be a linear algebraic group and let V be an H-module such that $k[V]^H$ is not finitely generated over k. (Examples were given in §8.) Then $k[GL(V)]^H$ cannot be finitely generated over k. For if it were, $k[V]^H$ would also be finitely generated over k by condition (b).

One of the purposes of these notes is to give a list of subgroups H of a reductive group G which satisfy the conditions in Theorem 9.3. The list will be seen to include maximal unipotent subgroups (Theorem 9.4), unipotent radicals of parabolic subgroups (Theorem 16.4), and subgroups H such that G/H has complexity ≤ 1 (Corollary 21.3). Both K. Pommerenning and V.L. Popov have conjectured that any subgroup of G which is normalized by a maximal torus satisfies the conditions in Theorem 9.3. The Popov-Pommerenning conjecture would be proven true if one could show it to hold for connected unipotent subgroups U which are normalized by a maximal torus T of G. Indeed, suppose that H is any closed connected subgroup normalized by T. Let V be the unipotent radical of H. Then $k[G/V]$ is finitely generated over k by assumption. But $k[G/H]$ consists of the invariants of (the reductive group) H/V on $k[G/V]$ and, so, is finitely generated by Theorem A. Thus, it suffices to show the Popov-Pommerenning conjecture holds for unipotent subgroups which are normalized by a maximal torus T of G. Let U be such a subgroup. Then $(TU)'' = G_1$ is reductive (Lemma 3.10). If U is a Grosshans subgroup of G_1, it is also a Grosshans subgroup of G as we see by applying Theorem 9.3(b) to the action of G_1 on G by right translation. Thus, we may always assume that TU is epimorphic in G.

Combining Theorems 9.3 and 5.6, we obtain one of the most important theorems in these notes.

Theorem 9.4. *Let G be a reductive group and let U be a maximal unipotent subgroup of G. Let A be a finitely generated, commutative k-algebra on which G acts rationally. Then A^U is finitely generated over k.*

Note. Theorem 9.4 was first proved in char k = 0 by Hadziev [44]; a representation-theoretic proof, valid in all characteristics and which also works for unipotent radicals of parabolic subgroups, may be found in [31]. The author came upon the idea of "codimension 2" embeddings and finite generation reading Nagata's exposition of Seshadri's proof of Weitzenböck's theorem [76; Chapter IV] and [98].

Corollary 9.5. *Let G be a reductive group, let U be a maximal unipotent subgroup of G, and let H be any closed subgroup of G which contains U. Let A be any finitely generated, commutative k-algebra on which G acts rationally. Then A^H is finitely generated over k.*
Proof. According to Theorem 9.3, we need only show that $k[G]^H$ is finitely generated. To do this, we may assume that H is connected (Theorem 4.1). Let $B = S{\cdot}U$ be a Borel subgroup of H, where S is a maximal torus in H. Then $k[G]^H = k[G]^B$ (by Lemma 1.1(vi)) $= (k[G]^U)^S$. But this algebra is finitely generated by Theorem 9.4 and Theorem A. QED

Corollary 9.6. *Let G be a reductive subgroup of a linear algebraic group L. Let H be a Grosshans subgroup of G. Let G_1 be any observable subgroup of L such that H is a subgroup of G_1. Then H is a Grosshans subgroup of G_1.*
Proof. Since H is observable in G and G is observable in L (Corollary 2.4), we see that H is observable in L (Corollary 2.3). Then, H is observable in G_1 since condition (2) of Theorem 2.1 holds in L and, so also in G_1. Now, G acts on L by right translation and so, by Theorem 9.3(b), the algebra $k[L]^H$ is finitely generated over k. This shows that H is a Grosshans subgroup of L. We now apply Corollary 4.5. QED

Corollary 9.7. *Let G be a reductive group and let H be a Grosshans subgroup of G. Let W be a finite-dimensional H-module. There is a finite-dimensional G-module V containing W as an H-submodule so that $k[V]^H$ is finitely generated over k.*
Proof. This is an immediate consequence of Theorems 2.1 (7) and 9.3 (b). QED

In a sense, Theorem 9.3 and its consequences are unsatisfactory in that the proof that A^H be finitely generated over k requires the action of a reductive group G (of which H is a subgroup) on A. However, the counter-examples to Hilbert's fourteenth problem strongly suggest that these results may be the best possible. For example, we saw in §8 that Annette A'Campo-Neuen showed that there is a linear action of $(\mathbf{G}_a)^{12}$ on $\mathbf{A}^{19}$ such that the algebra of invariants is not finitely generated. Since $(\mathbf{G}_a)^{12}$ is maximal unipotent in $SL(2,k) \times \ldots \times SL(2,k)$ (12-copies), it seems unlikely that the conditions in Theorem 9.3 may be weakened.

Let G be a reductive algebraic group and let G act rationally on a finitely generated, commutative k-algebra A. Let I be a G-invariant ideal in A. M. Nagata showed how the algebra $(A/I)^G$ is related to A^G. We state his result next and then apply it to study the relationship between $(A/I)^H$ and A^H. This result will be important in §14.

Theorem 9.8. [100; Lemma 2.4.7, p.23] *Let G be a reductive algebraic group. Let G act rationally on a finitely generated, commutative k-algebra R. Let I be an ideal in R which is invariant under G and let $f + I \in (R/I)^G$. Then there is a positive integer t and an element $F \in R^G$ such that $f^t - F \in I$.*

Note 1. The assumption that R be finitely generated is not needed. Indeed, the vector space $<G{\cdot}f>$ is finite dimensional since G acts rationally on R. Let R' be the k-subalgebra of R which is generated by $<G{\cdot}f>$ and let $I' = I \cap R'$. Then, $f + I' \in (R'/I')^G$ so, by what was just proved, there is an element $F \in (R')^G$ and a positive integer t so that $f^t - F \in I'$. Then, $F \in R^G$ and $f^t - F \in I$. QED

Note 2. If char $k = 0$, then we may take $t = 1$. Indeed, using the complete reducibility of G, we have $R = I \oplus V$, where V is a G-invariant subspace. Thus, $R/I = V$. In general, t is a power of the characteristic of k.

Theorem 9.9. *Let G be a reductive algebraic group acting rationally on a commutative k-algebra A. Let J be an ideal in A which is invariant under G. Let H be a closed subgroup. If $f + J \in (A/J)^H$, then there is a positive integer t and an $F \in A^H$ such that $f^t - F \in J$.*

Proof. The canonical mapping π from A to A/J gives rise to a homomorphism π' from $R = k[G/H] \otimes A$ to $S = k[G/H] \otimes (A/J)$ whose kernel is the G-invariant ideal $I = k[G/H] \otimes J$. Let φ: $R \to A$ (resp. φ': $S \to A/J$) be the algebra homomorphism defined by $\varphi(f \otimes a) = f(e)a$ (resp. $\varphi'(f \otimes (a + J)) = f(e)a + J$). Then $\pi \circ \varphi = \varphi' \circ \pi'$.

Let $f + J \in (A/J)^H$. According to Theorem 9.1, there is an element $f' \in S^G$ such that $\varphi'(f') = f + J$. We now apply Theorem 9.8 to see that there is an element $F' \in R^G$ and a positive integer t such that $(f')^t = F' + I$. We put $F = \varphi(F')$; then F is H-invariant by Theorem 9.1 and $\pi(F) = \pi(\varphi(F')) = \varphi'(\pi'(F')) = \varphi'(f'^t) = f^t + J$. QED

Exercises.

1. Let H and K be closed subgroups of G. Suppose that H (resp. K) is a Grosshans subgroup of K (resp. G). Show that H may not be a Grosshans subgroup of G.
2. Let H be a linear algebraic group. Let W be a finite-dimensional vector space with $H \subset GL(W)$. Corollary 9.7 raises the question as to whether there is a reductive group G such that (i) $k[G]^H$ is finitely generated over k and (ii) $H \subset G \subset GL(W)$. Show that such a G exists if and only if $k[GL(W)/H]$ is finitely generated over k.
3. Let G be reductive and let H be a Grosshans subgroup of G. Let X be the affine variety such that $k[X] = k[G/H]$ and let $x \in X$ be such that $G_x = H$ and $G/H \simeq G{\cdot}x$. Let Y be any affine variety on which G acts morphically. Show that the mapping φ from $k[Y \times X]$ to $k[Y]$ given by $(\varphi f)(y) = f(y,x)$ gives an isomorphism from $k[Y \times X]^G$ to $k[Y]^H$.
4. Let A be a commutative k-algebra on which G acts rationally. Let I and J be G-invariant radical ideals in A such that $I \subset J$. Show that $I = J$ if and only if $I^U = J^U$.

§10. The Theorems of Roberts and Weitzenböck

Let $G = SL(2,k)$ be the group consisting of all 2×2 matrices with entries in k and having determinant $= 1$. We denote the co-ordinate functions on G by x_{ij}, $1 \leq i,j \leq 2$. Let U (resp. U^-) be the subgroup of G consisting of all matrices (u_{ij}) such that $u_{11} = u_{22} = 1$ and $u_{21} = 0$ (resp. $u_{12} = 0$). Let T be the subgroup of G consisting of diagonal matrices. Then U (resp. T) is a maximal unipotent subgroup (resp. maximal torus) in G and $B = TU$ is a Borel subgroup of G. Let ω be the character of T defined by $\omega(t) = x_{11}(t)$ for all $t \in T$. The Borel subgroup B^- which is "opposite" to B is $B^- = TU^-$.

The group G acts on $\mathbf{A}^2$, the vector space consisting of all 2×1 matrices with entries in k, by $g{\cdot}v = gv$ for $g \in G$, $v \in \mathbf{A}^2$. Let x and y be the usual co-ordinate functions on $\mathbf{A}^2$. Let

$$v = \begin{pmatrix} 1 \\ 0 \end{pmatrix}.$$

Then, $G_v = U$ and $G{\cdot}v = \mathbf{A}^2 - \{0\}$. Now, consider the mapping $\varphi\colon k[G{\cdot}v] \to k[G]^U$ given by $(\varphi f)(g) = f(g{\cdot}v)$. Then $\varphi(x) = x_{11}$ and $\varphi(y) = x_{21}$. Since $k(G) = k(x_{11}, x_{12}, x_{21})$ is separable over $k(x_{11}, x_{21})$, we see that $G/U \simeq \mathbf{A}^2 - \{0\}$ (Theorem 1.11). In fact, according to Theorem 4.2, $k[G]^U \simeq k[\mathbf{A}^2] = k[x,y]$.

If $g \in G$, with say

$$g = \begin{pmatrix} a & b \\ c & d \end{pmatrix},$$

then $\ell_g x_{11} = dx_{11} - bx_{21}$ and $\ell_g x_{21} = -cx_{11} + ax_{21}$. Since φ is G-equivariant with respect to the natural action of G on $k[\mathbf{A}^2]$ and left translation on $k[G]$ (Exercise 3, §1), these equations hold with x_{11} replaced by x and x_{21} by y. In particular, if Z is any rational G-module, then by Theorem 9.1, $(k[x,y] \otimes Z)^G \simeq Z^U$. If $Z = A$ is a finitely generated, commutative algebra over k, then A^U is also finitely generated over k (Theorem 9.4).

Let V_d be the subspace of $k[x_{11}, x_{21}]$ consisting of all polynomials homogeneous of degree d and on which G acts by left translation. If $f \in V_d$ and $t \in T$, we see that $r_t f = (d\omega)(t)f$. Furthermore, according to Exercise 1, below, $\dim(V_d)^U = 1$; in fact, $(V_d)^U = \{cx_{21}{}^d : c \in k\}$, using the formulas for ℓ_g given above.

- From now on in this section we shall assume that char $k = 0$ unless otherwise noted.

We shall identify V_d with polynomials in $k[x,y]$ which are homogeneous of degree d. We may take as a basis of V_d the elements $w_i = (-1)^i \cdot (1/i!)x^iy^{d-i}$ for $i = 0,\ldots,d$. Let $u = u_b \in U$ with $u_{12} = b$. Then $u{\cdot}w_i = (-1)^i{\cdot}(1/i!)(x - by)^iy^{d-i} = \Sigma_r\, (1/(i-r)!)b^{i-r}w_r = w_i + bw_{i-1} + (b^2/2!)w_{i-2} + \ldots + (b^i/i!)w_0$.

The observations above give some information concerning invariants of the group $\mathbf{G}_a$. For a moment, let H be any algebraic group and let V be any finite-dimensional rational H-module. Is $k[V]^H$ finitely generated over k? (We know now that the answer, in general, is negative.) In 1932, R. Weitzenböck thought he had found a proof for an affirmative answer, valid for all H in case char $k = 0$. But Hermann Weyl, in reviewing the paper, found a gap. However, Weitzenböck's argument was valid for one-parameter groups. Since the case $H = \mathbf{G}_m$ is well understood [10; Lemma 8.19, p.123], Weitzenböck's proof is now taken as a theorem about the invariants of $\mathbf{G}_a$.

Theorem 10.1. *Suppose that char $k = 0$ and let $H = \mathbf{G}_a$. Let V be any finite-dimensional rational H-module. Then $k[V]^H$ is finitely generated over k.*

Proof. We use some of the notation introduced above taking $H = \mathbf{G}_a = U$ and $G =$

$SL(2,k)$. Suppose, for a moment, that we could show that the action of H on V extends to an action of G on V. Then, Weitzenböck's result would follow immediately from Theorem 9.4. Thus, we need only prove the following lemma.

Lemma 10.2. *Let char $k = 0$ and let V be a finite-dimensional $U = \mathbf{G}_a$ - module. Then the action of U on V extends to an action of $G = SL(2,k)$ on V.*

Proof. We shall identify U with a closed subgroup of $GL(W)$. The Lie algebra of U, denoted by $L(U)$, consists of all cX, where $c \in k$, and X is a nilpotent matrix. Furthermore, the mapping $L(U)$ to U given by $cX \rightarrow \exp(cX) = I + (cX) + (cX)^2/2! + \ldots$ is surjective [10; 7.3, p.106]. The matrix X can be brought to a direct sum of Jordan blocks, J_i, each of which has zeros on the diagonal. To prove the lemma, we may assume that there is only one such block, say J. Suppose that J is a $d \times d$ matrix. Then, $\exp(cJ)$ is the matrix of u_c in the basis $\{w_i\}$ of V_{d-1}, given above. Thus, the action of $\mathbf{G}_a$ on W is that of U on V_{d-1}, which is an $SL(2,\mathbf{C})$-module. QED

Corollary 10.3. *Let char $k = 0$. Let H be an affine algebraic group. Let $\mathfrak{R}_u H$ be the unipotent radical of H. Let V be a finite-dimensional H-module. If $\dim \mathfrak{R}_u H \leq 1$, then $k[V]^H$ is finitely generated over k.*

Proof. Let S be the algebra of $\mathfrak{R}_u H$-invariants in $k[V]$. According to Theorem 10.1, S is finitely generated over k. The reductive group $R = H/\mathfrak{R}_u H$ acts on S; by Theorem A, S^R is finitely generated over k. But $S^R = k[V]^H$. QED

Note. Suppose that char $k = p > 0$. Does Weitzenböck's theorem still hold, i.e., may the following question be answered in the affirmative? Let char $k = p > 0$ and let $H = \mathbf{G}_a$. Let V be a finite-dimensional H-module. Is $k[V]^H$ finitely generated over k? Of course, the conjecture is true if the action of H on V extends to an action of $G = SL(2,k)$ on V (by Theorem 9.4). However, not all such actions extend (Exercise 7) and the answer to the question is not known in general [34], [35].

In the rest of this section, we let $k = \mathbf{C}$ and describe some highlights of the invariant theory of binary forms, a subject which was studied intensively in the nineteenth century. We denote an element $f \in V_d$ by

$$a_0x^d + \binom{d}{1}a_1x^{d-1}y + \ldots + \binom{d}{i}a_ix^{d-i}y^i + \ldots + \binom{d}{d}a_dy^d$$

and usually think of the a_i as being co-ordinate functions on V_d.

The group $G = SL_2(\mathbf{C})$ acts on V_d as described before; our goal is the study of the algebras $\mathbf{C}[V_d]^U$ and $\mathbf{C}[V_d]^G$. A calculation shows that if $t \in T$, then $(t{\cdot}a_i)(f) = a_i(t^{-1}{\cdot}f) = \omega(t)^{d-2i}a_i(f)$, i.e., $t{\cdot}a_i = \omega(t)^{d-2i}a_i$. If $u_b =$

$$\begin{pmatrix} 1 & b \\ 0 & 1 \end{pmatrix},$$

then, $u_b{\cdot}a_i =$

$$\sum_{j=0}^{i} \binom{i}{j} a_j b^{i-j}.$$

Theorem 10.4 (M. Roberts, 1871). *$(C[x,y] \otimes C[V_d])^G = C[A^2 \times V_d]^G \simeq C[V_d]^U$.*
Proof. This immediate consequence of Theorem 9.1 was noted above. QED

In the nineteenth century, the invariants on the left-hand side of the isomorphism in Theorem 10.4 were called "covariants" and on the right-hand side "semi-invariants". Let us write a covariant as $\Sigma_i x^i y^{d-i} \otimes f_{d-i}$ where we are assuming that the elements in both $\mathbf{C}[x,y]$ and $\mathbf{C}[V_d]$ are homogenous. (This is harmless since the action of G preserves homogenous components.) The isomorphism between covariants and semi-invariants (given generally in the proof of Theorem 9.1) sends $\Sigma_i x^i y^{d-i} \otimes f_{d-i}$ to $\Sigma_i x^i y^{d-i}(e) f_{d-i}$ where e is the identity in G. Since $x(e) = x_{11}(e) = 1$ and $y(e) = x_{21}(e) = 0$, the covariant $\Sigma_i x^i y^{d-i} \otimes f_{d-i}$ is sent to f_0, its "leading coefficient."

Theorem 10.5. *Let $V_d' = \{f \in V_d : a_0(f) \neq 0\}$ and $W = \{f \in V_d : a_0(f) \neq 0,\ a_1(f) = 0\}$. The mapping φ: $U \times W \to V_d'$, $\varphi(u,w) = u{\cdot}w$, is an isomorphism.*
Proof. Let u_b be as above. Then, as we saw earlier, $u_b{\cdot}x = x - by$ and $u_b{\cdot}y = y$. Let $f \in V_d'$ and let $b = a_1(f)/a_0(f)$; we denote u_b by u_f. Then $u_f{\cdot}f \in W$ and $\varphi(u_f^{-1}, u_f{\cdot}f) = f$. Thus, φ is surjective. It is injective since $u{\cdot}w \in W$, for $w \neq 0$, if and only if $u = e$. Finally, the calculation above shows that φ has an inverse. QED

Now, let us look at the U-invariants in $\mathbf{C}[V_d'] = \mathbf{C}[V_d][1/a_0]$. Since a_0 is U-invariant, we have $(\mathbf{C}[V_d][1/a_0])^U = \mathbf{C}[V_d]^U[1/a_0]$. Also, U acts on $U \times W$ via $u{\cdot}(u_1, w) = (uu_1, w)$ and φ is U-equivariant with respect to this action. Thus, $\mathbf{C}[V_d']^U \simeq \mathbf{C}[W] = \mathbf{C}[a_0, a_2, \ldots, a_d][1/a_0]$. The generators of $\mathbf{C}[V_d']^U$ are given by $f \in V_d' \to u_f{\cdot}f \to a_i(u_f{\cdot}f)$, a rational function which may be written as $S_i(a_0, a_1, \ldots, a_d)/a_0^{i-1}$. In particular, we may use the calculation given above for $u_b{\cdot}a_i$ to see that

$$\begin{aligned}
S_0 &= a_0,\\
S_2 &= a_0 a_2 - a_1^2,\\
S_3 &= a_0^2 a_3 - 3a_0 a_1 a_2 + 2a_1^3,\\
S_4 &= a_0^3 a_4 - 4a_0^2 a_1 a_3 + 6a_0 a_1^2 a_2 - 3a_1^4.\\
&\ldots
\end{aligned}$$

Theorem 10.6. *$C[V_d']^U = C[S_0, S_2, \ldots, S_d][1/a_0]$ where the S_i are as above and are algebraically independent.*
Proof. We need only prove the last statement. But this follows from the form of the S_i since for each $i = 2, \ldots, d$, a_i appears first in S_i. QED

For small d, namely, $d = 1,2,3,4$, we can go further than the theorem and actually calculate $\mathbf{C}[V_d]^U$ and $\mathbf{C}[V_d]^G$. (Except for the case $d = 2$, we leave the calculation of the G-invariants as an exercise.)

d = 1: *$C[V_1]^U = C[a_0]$ and $C[V_1]^G = C$.*

d = 2: *$C[V_2]^U = C[a_0, S_2]$ and $C[V_2]^G = C[S_2]$.*

Proof. According to Theorem 10.6, every U-invariant $S(a_0,a_1,a_2)$ may be written in the form $P(a_0,S_2)/a_0^e$ where P is some polynomial. Now, suppose that $a_0^eS(a_0,a_1,a_2) =$

$$P(a_0,S_2) = b_0(S_2)a_0^m + b_1(S_2)a_0^{m-1} + \dots + b_m(S_2),$$

where $b_m(S_2) \neq 0$ and $e \geq 1$. Let $a_0 = 0$. Then, using the explicit expressions for the S_i given above, we see that $0 = b_m(-a_1^2)$. But this is impossible since a_1 is transcendental over k unless $b_m = 0$. This contradiction proves the statement about U-invariants. The G-invariants are as described since S_2 is T-invariant and $C[V_1]^G = (C[V_1]^U)^T$. QED

d = 3: *$C[V_3]^U = C[a_0, S_2, S_3, D]$ and $C[V_3]^G = C[D]$ where $D = a_0^2a_3^2 - 6a_0a_1a_2a_3 + 4a_0a_2^3 + 4a_1^3a_3 - 3a_1^2a_2^2$ is the discriminant of a cubic form and $4S_2^3 + S_3^2 - a_0^2D = 0$.*

Proof. For $i = 2,3$, let $P_i(a_1,a_2,a_3) = S_i(0,a_1,a_2,a_3)$. Then, $P_2 = -a_1^2$ and $P_3 = 2a_1^3$ so that $4P_2^3 + P_3^2 = 0$ which implies that a_0 divides $4S_2^3 + S_3^2$. In fact, $4S_2^3 + S_3^2 = a_0^2D$ where D is as above. The polynomial D is U-invariant as was just shown; a quick calculation shows it is also T-invariant. Thus, it is G-invariant. By means of the preceding equation, we may solve for S_3^e, $e \geq 2$, and conclude from Theorem 10.6 that any U-invariant $S(a_0,a_1,a_2,a_3)$ has the form $P(a_0,S_2,S_3,D)/a_0^e$ where P has degree 0 or 1 in S_3. In particular, suppose that

$$a_0^eS(a_0,a_1,a_2,a_3) = b_0(S_2,S_3,D)a_0^m + \dots + b_m(S_2,S_3,D)$$

where $e \geq 1$ and $b_m(S_2,S_3,D) \neq 0$. (We seek a contradiction.) Putting $a_0 = 0$ in this equation gives (*) $0 = b_m(-a_1^2, 2a_1^3, 4a_1^3a_3 - 3a_1^2a_2^2)$. Now, D cannot occur in b_m since a_3 is algebraically independent of a_1 and a_2. Also, S_3 cannot occur (to the first power in b_m) since $2a_1^3$ cannot be canceled by any power of $-a_1^2$. Then, $b_m(-a_1^2) = 0$ implies that $b_m = 0$, which is the desired contradiction. QED

d = 4: *$C[V_4]^U = C[a_0, S_2, S_3, I, J]$ and $C[V_3]^G = C[I, J]$ where $I = a_0a_4 - 4a_1a_3 + 3a_2^2$, $J = a_0a_2a_4 - a_0a_3^2 + 2a_1a_2a_3 - a_1^2a_4 - a_2^3$ and $a_0^3J - a_0^2S_2I + 4S_2^3 + S_3^2 = 0$.*

Proof. We proceed as in the case $d = 3$. For $i = 2,3,4$, let $P_i(a_1,a_2,a_3,a_4) = S_i(0,a_1,a_2,a_3,a_4)$. Thus, $P_2 = -a_1^2$, $P_3 = 2a_1^3$, and $P_4 = -3a_1^4$. Then, $3P_2^2 + P_4 = 0$ which implies that a_0 divides $3S_2^2 + S_4$. In fact, $3S_2^2 + S_4 = a_0^2I$ where I is as above. Using this equation, we may always replace S_4 by $a_0^2I - 3S_2^2$. Now, we saw in the case $d = 3$ that $4S_2^3 + S_3^2 = a_0^2D$ where $D = a_0^2a_3^2 - 6a_0a_1a_2a_3 + 4a_0a_2^3 + 4a_1^3a_3 - 3a_1^2a_2^2$. Putting $a_0 = 0$ in this expression for D, we see that $D \equiv -a_1^2I \equiv P_2I \pmod{a_0}$. Then, $S_2I - D = a_0J$ where J is as above. The two equations $S_2I - D = a_0J$ and $4S_2^3 + S_3^2 = a_0^2D$ give the relation $a_0^3J - a_0^2S_2I + 4S_2^3 + S_3^2 = 0$. This equation also enables us to solve for S_3^2 and then conclude from Theorem 10.6 that any U-invariant $S(a_0,a_1,a_2,a_3,a_4)$ has the form $P(a_0,S_2,S_3,I,J)/a_0^e$ where P has degree 0 or 1 in S_3. In particular, suppose that

$$a_0^eS(a_0,a_1,a_2,a_3,a_4) = b_0(S_2,S_3,I,J)a_0^m + \dots + b_m(S_2,S_3,I,J)$$

where $e \geq 1$ and $b_m(S_2,S_3,I,J) \neq 0$. We may now proceed as in the case $d = 3$ to complete the proof (Exercise 5).

Note 1. Hilbert was actually able to describe explicitly the generators for $\mathbf{C}[V_d]^U$ and $\mathbf{C}[V_d]^G$ up to integral closure. We shall sketch his ideas here. Let $n = [d/2]$. We define functions $C_2, C_4, \ldots, C_{2n} \in \mathbf{C}[V_d]$ by $C_{2m} =$

$$\sum_{i=0}^{m-1} (-1)^i \binom{2m}{i} a_i a_{2m-i} + (1/2)(-1)^m \binom{2m}{m} a_m^2 .$$

Each of the C_i is U-invariant (Exercise 4). Let $C_0 = a_0$ and let $F_0, F_2, \ldots, F_{2n}$ be the corresponding covariants (i.e., as in Theorem 10.4). A covariant may be considered as a polynomial in x and y. Then, replacing the F_i's by suitable powers, we may compute the resultant of F_0 and $u_2F_2 + \ldots + u_{2n}F_{2n}$, where the u_i are indeterminates. In the resultant expansion, the coefficients of the various monomials in the u's are actually G-invariants, say $f_1, \ldots, f_r$. The following results are proved in [33].

- $\mathbf{C}[V_d]^G$ is integral over $\mathbf{C}[f_1, \ldots, f_r]$ (Exercise 8, p.208).

- $\mathbf{C}[V_d]^U$ is integral over $\mathbf{C}[V_d]^G[C_0, \ldots, C_{2n}]$ (Exercise 33, p.228).

Note 2. Theorem 10.5 may be viewed as a special case of the following result [15; Theorem 1.4]. Let char $k = 0$ and let G be a connected reductive group. Let Z be a normal G-variety and let $z \in Z$ be such that G_z is a parabolic subgroup of G. Let P be the parabolic subgroup of G which is opposite to G_z and let $L = P \cap G_z$. Then, there exists a locally closed affine subvariety W of Z such that (1) W contains z and is invariant under L; (2) $\mathfrak{R}_uP{\cdot}W$ is open in Z; (3) the mapping of $\mathfrak{R}_uP \times W \to \mathfrak{R}_uP{\cdot}W$, given by $(p,w) \to p{\cdot}w$, is an isomorphism. Using this, one may extend Theorem 10.6 to irreducible representations of G [41].

Exercises.

1. Show that $\dim(V_d)^U = 1$. (Hint: use the fact that U^-TU is open in G.)
2. Suppose that char $k = 0$. Show that the V_d, $d \geq 0$, give all the irreducible finite-dimensional representations of G.
3. Show that the locally nilpotent derivation associated to the action of U on $\mathbf{C}[V_d]$ is

$$\Omega = a_0\partial/\partial a_1 + 2a_1\partial/\partial a_2 + \ldots da_{d-1}\partial/\partial a_d .$$

4. Using Exercise 3, show that the C_i are U-invariant.
5. Complete the argument when $d = 4$.
6. Show that the G-invariants in the cases $d = 2,3,4$ are as described.
7. We give here an example of an action of U on a finite-dimensional vector space V which does not extend to an action of G on V. To do this, it is enough to display a finite-dimensional representation of U which is not injective. (Why?) Let char $k = 2$. (a) Let $V = \mathbf{A}^2$ and let U act on V via the matrix representation

$$\begin{pmatrix} 1 & a^2+a^4 \\ 0 & 1 \end{pmatrix}$$

for $a \in k$. Show that this gives a representation of U on V which does not extend to a representation of G on V.
(b) Show that $k[V]^U$ is finitely generated over k.
(c) Show that this representation of U extends to a representation of G on a vector space having dimension ≤ 20.
(d) Consider the representation of U coming from the matrix representation

$$\begin{pmatrix} 1 & a^2+a^4 \\ 0 & 1 \end{pmatrix} \oplus \begin{pmatrix} 1 & a^2 \\ 0 & 1 \end{pmatrix} \oplus \begin{pmatrix} 1 & a^4 \\ 0 & 1 \end{pmatrix}.$$

(i) Does this (faithful) representation of U on $V \oplus V \oplus V$ extend to a representation of G on $V \oplus V \oplus V$?
(ii)* Is $k[V \oplus V \oplus V]^U$ finitely generated over k?

§11. Geometric Examples

A. Multiplicity free actions

In this section, we shall look at some examples of the transfer principle which were of great importance in the development of invariant theory and in its applications to physics. Each of the situations considered below (except Example E) is an example of a "spherical subgroup" so we begin by explaining this important concept.

Definition. Let G be an affine algebraic group, let B be a Borel subgroup of G, and let $X(B)$ be the character group of B. Let Z be a rational G-module. The action of G on Z is *multiplicity-free* if for each $\chi \in X(B)$, the subspace $Z_\chi = \{z \in Z : b\cdot z = \chi(b)z$ for all $b \in B\}$ is of dimension ≤ 1. A subgroup H of a connected group G is said to be *spherical* if the action of G by left translation on $k[G/H]$ is multiplicity-free.

Theorem 11.1. *Let G act morphically on an irreducible quasi-affine variety X. The action of G on k[X] is multiplicity-free if and only if there is a Borel subgroup B of G and a point $x \in X$ so that the orbit $B\cdot x$ is open in X.*

In this context, we shall say that B has an "open orbit" on X. Since the Borel subgroups of G are conjugate [10; Theorem 11.1, p.147], if some Borel subgroup has an open orbit on X, every Borel subgroup does. In this section, we shall only use the sufficiency of the condition in Theorem 11.1, which is easy to prove. The proof of necessity will be given in Lemma 19.7(b). So suppose that the orbit $B\cdot x$ is open in X. Let $\chi \in X(B)$ and suppose that there are non-zero functions $z_1, z_2 \in k[X]$ such that $b\cdot z_i = \chi(b)z_i$ for all $b \in B$. Then, $z_1/z_2 \in k(X)^B$; but this field is k since $B\cdot x$ is open in X.

Examples. *Let G be connected, reductive and let $B = TU$ be a Borel subgroup of G.*

(1) *Let U be a maximal unipotent subgroup of G. Then U is a spherical subgroup of G.*
Proof. Let B^- be the Borel subgroup "opposite" to B. According to the Bruhat decomposition of G, the variety B^-U is open in G so the orbit $B^-{\cdot}eU$ is open in G/U. QED

(2) *Let H be any closed subgroup of G which contains U. Then H is spherical.*
Proof. The orbit $B^-{\cdot}eH$, being the image of B^-U, is open in G/H. QED

(3) *The subgroup $G_1 = \{(g,g) : g \in G\}$ is spherical in $G \times G$.*
Proof. (Exercise 1).

(4) *Let V be a finite-dimensional rational G-module and let $v \in V$ be a (non-zero) highest weight vector, i.e., there is a character $\omega \in X(T)$ such that $tu{\cdot}v = \omega(t)v$ for all $t \in T$, $u \in U$. Then $cl(G{\cdot}v) = G{\cdot}v \cup \{0\}$ and G_v is a spherical, Grosshans subgroup of G.*
Proof. Applying Corollary 9.5, we see that G_v is a Grosshans subgroup of G; it is a spherical subgroup of G according to Example (2). QED

The next theorem really is an argument about the complexity of an action and we shall prove it later in that context (Theorem 20.2).

Theorem 11.2. *Let $H = TU$ be a connected, solvable linear algebraic group where T is a maximal torus in H and $U = \mathfrak{R}_uH$. Let H act morphically on an affine variety X. Suppose that $k[X]$ is a unique factorization domain and that the set of units in $k[X]$ is k^*. Suppose also that there are functions $f_1, \ldots, f_m$ and linearly independent characters $\chi_i \in X(T)$ such that (i) $k(X)^U = k(f_1, \ldots, f_m)$ and (ii) $t{\cdot}f_i = \chi_i(t)f_i$ for all $t \in T$. Then, $k[X]^U$ is a finitely generated polynomial algebra over k and the action of H on $k[X]$ is multiplicity-free. In fact, let $f_i = a_i/b_i$ where the a_i, b_i are T-weight functions in $k[X]^U$; then, the generators of $k[X]^U$ are the irreducible factors of the a_i and b_i.*
Proof. Let $f \in k(X)^U$ and let $\chi \in X(T)$ be such that $t{\cdot}f = \chi(t)f$ for all $t \in T$. We shall show that $f = a/b$ where $a,b \in k[X]^U$ and are T-weight vectors. Indeed, there are polynomials $a',b' \in k[X]$ so that $f = a'/b'$. Let $V = \{c \in k[X] : cf \in k[X]\}$. Then V is non-zero and H-invariant. Therefore, there is a non-zero eigenvector of H in V, say, b. Let $a = bf$.

As was just shown, we may write f_i as a_i/b_i where a_i, $b_i \in k[X]^U$ and are T-weight functions. Let p_j, $1 \leq j \leq r$, be all the irreducible polynomials appearing in the factorizations of the a_i's and b_i's. The p_j are H-eigenfunctions (Exercise 2).

Let $f \in k[X]^U$ and let $\chi \in X(T)$ be such that $t{\cdot}f = \chi(t)f$ for all $t \in T$. There are polynomials $a(X_1, \ldots, X_m)$, $b(X_1, \ldots, X_m)$ in indeterminates $X_1, \ldots, X_m$ such that $f = a(f_1, \ldots, f_m)/b(f_1, \ldots, f_m)$. Thus, (*) $b(f_1, \ldots, f_m)f = a(f_1, \ldots, f_m)$. Let $cf_1^{e(1)} \ldots f_m^{e(m)}f$ be a non-zero term appearing on the left-hand side of (*). Since the χ_i's are linearly independent over $\mathbf{Z}$, this is the only such term with T-weight $e(1)\chi_1 + \ldots + e(m)\chi_m + \chi$. There must be one and only one term $f_1^{d(1)} \ldots f_m^{d(m)}$ on the right-hand side of (*) having the same weight. Thus, (**) $f_1^{e(1)} \ldots f_m^{e(m)}f = c'f_1^{d(1)} \ldots f_m^{d(m)}$ for some $c' \in k^*$. Clearing out the denominators and comparing prime factorizations on both sides of (**), we see that f must be a monomial in the p_j's. This shows that $k[X]^U$ is finitely generated over k.

Let $k[X]^U_\chi = \{f \in k[X]^U : t{\cdot}f = \chi(t)f$ for all $t \in T\}$. Equation (**) shows that $\dim k[X]^U_\chi \leq 1$, i.e., the action of H on $k[X]$ is multiplicity-free. Indeed, (**) shows that any $f \in k[X]^U_\chi$ has the form $f_1^{a(1)} \ldots f_m^{a(m)}$, where $a(1), \ldots, a(m)$ are the unique integers such that $a(1)\chi_1 + \ldots + a(m)\chi_m = \chi$.

Now, let us show that the p_i's are algebraically independent over k. Each p_j is a T-weight function since the a_i and b_i are. Suppose that $t{\cdot}p_j = \mu_j(t)p_j$ for some $\mu_j \in X(T)$. Let $d(1), \ldots, d(r)$ be non-negative integers. The monomial $p_1^{d(1)} \ldots p_r^{d(r)}$ has T-weight $\chi = d(1)\mu_1 + \ldots + d(r)\mu_r$; it is the only monomial in the p_i's with weight χ since the p_i's are irreducible and $\dim k[X]^U_\chi \leq 1$. It follows that the p_i's are algebraically independent since in any relation $\Sigma c p_1^{e(1)} \ldots p_r^{e(r)} = 0$, the different monomials all have different T-weights. QED

From now on in this section, we shall assume that G is a connected, reductive algebraic group with Borel subgroup $B = TU$. In case $G = SL(n,k)$, we let $B = TU$ where T (resp. U) is the subgroup of G consisting of diagonal matrices (resp. upper triangular matrices with 1's on the diagonal). Also, for $1 \leq i \leq n - 1$, let ω_i be the character of T given by $\omega_i(t_{ij}) = t_{11}\ldots t_{ii}$. The transpose of a matrix m will be denoted by m^T.

The first lemma below is a technical one which allows us to state our later results in the traditional language. The second result covers several key examples.

Lemma 11.3. *Let char $k = 0$. Let H be a subgroup of G. Let V be a finite-dimensional G-module. Let $v \in V$ be an element such that $H = G_v$, each irreducible component of $cl(G{\cdot}v) - G{\cdot}v$ has codimension ≥ 2 in $cl(G{\cdot}v)$, and $cl(G{\cdot}v)$ is normal. Let Z be any rational G-module. Then the mapping $(k[V] \otimes Z)^G$ to Z^H given by $\Sigma_i f_i \otimes z_i \rightarrow \Sigma_i f_i(v)z_i$ is surjective.*

Proof. Let $X = cl(G{\cdot}v)$. Since X is normal, we have $k[X] = k[G{\cdot}v] = k[G/H]$ using Theorem 4.2. By the transfer principle, the mapping from $(k[X] \otimes Z)^G$ to Z^H given by $\Sigma_i f_i \otimes z_i \rightarrow \Sigma_i f_i(v)z_i$ is an isomorphism. Thus, we need only show that the restriction mapping from $(k[V] \otimes Z)^G$ to $(k[X] \otimes Z)^G$ is surjective. We may assume that Z has finite dimension. If W_1 and W_2 are finite-dimensional G-modules and if $p: W_1 \rightarrow W_2$ is a surjective G-equivariant linear mapping, then $p(W_1^G) = W_2^G$. For, applying complete reducibility, we see that there is a G-invariant subspace W of W_1, with $W_1 =$ (kernel p) $\oplus\, W$. Then $W^G \cong W_2^G$. QED

Theorem 11.4. *Let char $k = 0$. Let V be a finite-dimensional irreducible G-module. Let $v \in V$ be a highest weight vector with $b{\cdot}v = \omega(b)v$ for all $b \in B$. Let $H = G_v$ and let $X = cl(G{\cdot}v)$. Then H is a spherical, Grosshans subgroup of G. For $n = 0,1,2,\ldots$, let $E(n\iota\omega) = \{f \in k[G]^U : r_t f = \omega(t)^n f$ for all $t \in T\}$. Then, $k[G/H] = k[X] = \oplus_{n\geq 0}E(n\iota\omega)$. Let Z be any rational G-module. Then the mapping from $(k[V] \otimes Z)^G$ to Z^H given by $\Sigma_i f_i \otimes z_i \rightarrow \Sigma_i f_i(v)z_i$ is surjective.*

Proof. First, $G{\cdot}v \cong G/H$ by Theorem 1.11. We now show that X is normal. Let V^* be the dual space of V. Let $\varphi: V^* \rightarrow k[G]$ be defined by $(\varphi\mu)(g) = \mu(g{\cdot}v)$ for all $\mu \in V^*$ and $g \in G$. The mapping φ is non-zero and G-equivariant, where G acts on $k[G]$ by left translation (Exercise 3, §1); hence, $\varphi(V^*)$ is a non-zero subspace of $E(\iota\omega)$. Since $E(\iota\omega)$ is irreducible (by Exercise 4), we see that the image of V^* under φ is $E(\iota\omega)$. Thus, the mapping φ gives rise to an injective map from $k[X]$ to $k[G]^H$ whose image (by the calculation above) contains all the $E(n\omega)$, $n = 0,1,\ldots$. Let $T_1 = \{t \in T : \omega(t) = 1\}$; if $\chi \in X(T)$, then $\chi \mid T_1 = 0$ if and only if $\chi = m\omega$ for some integer m. Since H contains T_1U, we have $\varphi k[X] \supset \oplus E(n\omega) = k[G/T_1U] \supset k[G/H] \supset \varphi k[X]$. Thus, $k[X] = k[G/H]$ and X is normal by Lemma 1.8; the other statements in the Theorem follow from Lemma 11.3 and Example 4. QED

Before getting to the examples, we shall indicate how the concept of multiplicity-free actions is relevant to the construction of invariants.

Lemma 11.5. *Let char $k = 0$. Let V_1 and V_2 be finite-dimensional, irreducible G-modules.*
(a) *There is a non-zero, G-invariant bilinear form on $V_1 \times V_2$ if and only if $V_2 = V_1^*$, the dual space of V_1; in this case, the G-invariant bilinear form is unique up to scalar multiple.*
(b) *There is a non-zero, G-invariant element in $V_1 \otimes V_2$ if and only if $V_2 = V_1^*$; also, $\dim(V_1 \otimes V_1^*)^G = 1$.*
Proof. To prove necessity in (a), let $<v_1, v_2>$ be a non-zero G-invariant bilinear form on $V_1 \times V_2$. The mapping $\varphi: V_1 \to V_2^*$ given by $\varphi(v_1)(v_2) = <v_1, v_2>$ is non-zero and G-equivariant. Since both V_1 and V_2 are irreducible G-modules, φ is an isomorphism. If $V_2 = V_1^*$, the natural pairing gives a G-invariant bilinear form on $V_1 \times V_1^*$. Also, as we have just seen, there is a natural injection from the space of G-invariant bilinear forms on $V_1 \times V_1^*$ to the G-equivariant linear maps from V_1 to $V_1^{**} = V_1$. But according to Schur's lemma, this latter space has dimension 1. To prove necessity in (b), we note that a non-zero G-invariant element in $V_1 \otimes V_2$ gives rise to a non-zero, G-invariant linear functional on $(V_1 \otimes V_2)^* = V_1^* \otimes V_2^*$. But this corresponds to a non-zero, G-invariant bilinear mapping on $V_1^* \times V_2^*$. We now may apply statement (a). The sufficiency in each statement is left as an exercise (Exercise 3). QED

Now, let char $k = 0$. We shall denote the finite-dimensional, irreducible G-module having highest weight ω with respect to B by $V(\omega)$. Let Z be a rational G-module with, say, $Z = \oplus_i V(\omega_i)$. Let H be a closed subgroup of G with, say, $k[G/H] = \oplus_j V(\omega_j')$ where G acts by left translation. By the transfer principle, we have $Z^H \simeq (k[G/H] \otimes Z)^G = \Sigma_{i,j}(V(\omega_j') \otimes V(\omega_i))^G$. According to Lemma 11.5, $\dim (V(\omega_j') \otimes V(\omega_i))^G \leq 1$ and is equal to 1 if and only if $V(\omega_i) = V(\omega_j')^*$. In case Z is multiplicity-free and H is a spherical subgroup of G, it may be possible to give an explicit description of Z^H using this pairing. This approach has been extensively developed by Roger Howe who, e.g., has shown that most known cases of the famous First Fundamental Theorem may be derived in this way [51].

B. Affine geometry

> *As far as invariants go, we should then be justified in treating the affine space as the projective space with an absolute plane.*
> [116, p.255]

Let $G = SL(n,\mathbf{C})$ and let H be the subgroup of G characteristic of affine geometry, namely, the set of all matrices of the form

$$\begin{pmatrix} A & b \\ 0 & 1 \end{pmatrix}$$

where $A \in SL(n-1,\mathbf{C})$ and b is an $(n-1) \times 1$ column. Let V be the set of all $1 \times n$ row matrices over $\mathbf{C}$, i.e., $V = \{(a_1,\ldots,a_n) : a_i \in \mathbf{C}\}$, and let $v \in V$, $v = (0,\ldots,0,1)$. The group G acts on V by $g{\cdot}r = rg^{-1}$ for all $g \in G$ and $r \in V$. It is easy to check that $H = G_v$ and that v, in fact, is a highest weight vector corresponding to the character ω_{n-1}. By Theorem 1.11, $G/H \simeq G{\cdot}v = V - \{0\}$. Furthermore, since V is a normal variety, we have $\mathbf{C}[G]^H = \mathbf{C}[V]$ by Theorem 4.2. We apply Theorem 9.1 to obtain the following result: if Z is any rational G-module, then $(\mathbf{C}[V] \otimes Z)^G \simeq Z^H$. This has a more geometric flavor if we take the case $Z = \mathbf{C}[X]$, where X is an affine variety on which G acts morphically.

Theorem 11.6. *Let G, H and V be as above. Then H is a spherical, Grosshans subgroup of G and $k[G/H] = k[X] = \oplus_{n\geq 0}E(n\omega_{n-1})$. If X is any affine variety on which G acts morphically, then $C[V \times X]^G = (C[V] \otimes C[X])^G = C[X]^H$.*

In the isomorphism of Theorem 11.6, a G-invariant $\Sigma_i f_i \otimes f_i{}'$ in $\mathbf{C}[V] \otimes \mathbf{C}[X]$ is sent to an H-invariant on X by $\Sigma_i f_i \otimes f_i{}' \to \Sigma_i f_i(v) f_i{}'$. Thus, the H-invariants on X come from G-invariants on X when the "plane at infinity" is adjoined.

To obtain a version of Theorem 11.6 valid for $\mathbf{R}$, we may apply Theorem 9.2. (In the future, we usually will not give such analogues.)

Theorem 11.7. *Let X be an affine variety on which G acts morphically. Suppose that both X and the action of G on X are defined over $\mathbf{R}$. Then $(R[V] \otimes_R R[X])^{SL(n,R)} \simeq R[X]^{H(R)}$.*

> *... an invariant $J(uv\ldots)$ of a number of ground forms uv in the affine space "of rank 2" is derived by the specialization from a projective invariant involving an arbitrary anti-symmetric tensor...* . [116, p.256]

This example of Hermann Weyl, "linear manifold as absolute", is a special case of Theorem 11.4. In fact, let V be the n-dimensional vector space over $\mathbf{C}$ consisting of $n \times 1$ column matrices. Let e_i be the element in V having 1 in the ith row and 0's elsewhere. Then $\{e_1, e_2, \ldots, e_n\}$ is a basis for V; using it, we identity $SL(V)$ and $SL(n,\mathbf{C})$. We take B to be the subgroup of $SL(n,\mathbf{C})$ consisting of all upper triangular matrices. In the context of the theorem, we take $G: = SL(n,\mathbf{C})$, $B: =$ upper triangular matrices in G, $V: = \Lambda^2(V)$, and $v: = e_1 \wedge e_2$.

C. Invariants of the orthogonal group

> *Through relativity theory one has become thoroughly familiar with*

treating orthogonal vector invariants as affine invariants after adjoining the metric ground form... . [116; p.256]

Throughout this example, we shall assume that char $k \neq 2$. Let $G = SL(n,k)$, let S_n be the set of all $n \times n$ symmetric matrices over k, and let G act on S_n by $g{\cdot}s = gsg^T$ for all $g \in G$, $s \in S_n$. Let I be the $n \times n$ identity matrix. Some elementary linear algebra shows that the orbit $G{\cdot}I$ is closed and, in fact, is $\{s \in S_n : \det(s) = 1\}$. Let H be the special orthogonal group, i.e., $H = \{g \in G : gIg^T = I\}$. We wish to describe the algebra of U-invariants on G/H. To this end, we define polynomials $D_i \in k[S_n]$, $1 \leq i \leq n$, as follows. Let $s \in S_n$ and let s' be the $(n-i+1) \times (n-i+1)$ symmetric matrix obtained from s by deleting its first i-1 rows and i-1 columns; then $D_i(s) = \det(s')$. A direct computation shows that D_i is U-invariant and has T-weight $2\omega_{i-1}$. Also, since D_i is the determinant of a symmetric matrix, it is irreducible (Exercise 8).

Theorem 11.8. *$k[S_n]^U = k[D_1, \ldots, D_n]$, a polynomial algebra.*
Proof. Let S' consist of all diagonal matrices in S_n having non-zero determinant. The mapping $U \times S' \to S_n$, $(u,s) \to usu^T$, is injective so the dimension of its image is $\dim U + \dim S' = \dim S_n$. Furthermore, the image, itself, Y is open in S_n according to Theorem 5.1. Now, if $s = (s_{ij}) \in S'$, then $s_{ii} = D_i(s)/D_{i+1}(s)$ for $1 \leq i \leq n - 1$ and $s_{nn} = D_n(s)$ so $k(S') = k(D_1 \,|\, S', \ldots, D_n \,|\, S')$. It follows that if $f \in k(S_n)^U$, then $f|\,S'$ is a rational function in the $D_i \,|\, S'$. Thus, f itself is a rational function in the D_i and $k(S_n)^U = k(D_1, \ldots, D_n)$. We now apply Theorem 11.2 to the group $H = T_1U$, where T_1 is the group of all non-singular $n \times n$ diagonal matrices, to see that $k[S_n]^U = k[D_1, \ldots, D_n]$. Since the $D_i \,|\, S'$ are algebraically independent, so are the D_i. QED

Theorem 11.9. *Let H and G be as above. Then H is a spherical, Grosshans subgroup of G. The algebra of U-invariants on G/H is $k[D_2, \ldots, D_n]$, a polynomial algebra over k. If char $k = 0$ and if Z is any rational G-module, then the mapping $(C[S_n] \otimes Z)^G$ to Z^H given by $\Sigma_i f_i \otimes z_i \to \Sigma_i f_i(I) z_i$ is surjective.*
Proof. First, let $\pi: G \to S_n$ be given by $\pi(g) = g{\cdot}I = g{\cdot}g^T$. Let $X = G{\cdot}I = \{s \in S_n : \det s = 1\}$; X is a smooth affine variety. The mapping $d\pi$ sends the tangent space at the identity of G onto the tangent space of X at I so $G{\cdot}I$ is G/H according to Theorem 1.11. A calculation shows that the stabilizer in B of I is finite. Then, a dimension count shows that the orbit $B{\cdot}I$ is open in X so H is a spherical subgroup of G; since H is semi-simple, it is also a Grosshans subgroup of G (Corollary 4.6). The restriction map from $k[S_n]$ to $k[X]$ sends $k[S_n]^U$ to $R = k[D_2 \,|\, X, \ldots D_n \,|\, X]$ and $k[X]^U$ is integral over R according to Theorem 9.9. Let S'' consist of all diagonal matrices in S_n having determinant 1. The image of the mapping $U \times S'' \to X$, $(u,s) \to usu^T$, is open in X. Furthermore, we see that R is a polynomial algebra since for $s = (s_{ij}) \in S''$, we have $D_i(s) = s_{ii} \ldots s_{nn}$ for $2 \leq i \leq n - 1$ and $D_n(s) = s_{nn}$. Then, reasoning as in the proof of Theorem 11.8, we see that $k(X)^U = k(D_2 \,|\, X, \ldots D_n \,|\, X)$. Since R is integrally closed in its quotient field, we see that any function in $k[X]^U$ is in R. QED

Note. Let $k = \mathbf{C}$ and let $V = \mathbf{A}^n$. Often, in the applications to physics, Z is taken to be an H-invariant subspace of $W = V \otimes \ldots \otimes V \otimes V^* \otimes \ldots \otimes V^*$. The space S_n is taken to be a subspace of $V^* \otimes V^*$. The multilinear G-invariants on $W \otimes V^* \otimes V^*$ are generated by $<v, \lambda>$, the usual pairing on $V \times V^*$, and the various determinants det[v_1, ... , v_n], det[λ_1, ... , λ_n]. The invariants of H on Z are then calculated by restriction (i.e., "contraction" with the elements in Z). Indeed, if F is a G-invariant on $W \otimes V^* \otimes V^*$, then the corresponding H-invariant is $f(w) = F(w,s_o)$ where s_o is the symmetric matrix defining the particular orthogonal group (Exercise 3, §9). Some of these calculations may now be performed by computers [2].

The result above may also be restated for the symplectic group. (Since the proofs are similar to those for the orthogonal group, we place them in Exercise 9.) Let $G = SL(2n,k)$, let S_{2n} be the set of all $2n \times 2n$ skew-symmetric matrices over k, and let G act on S_{2n} by $g \cdot s = gsg^T$ for all $g \in G$, $s \in S_{2n}$. Let $J \in S_{2n}$ be the $2n \times 2n$ matrix with

$$\begin{pmatrix} 0 & 1 \\ -1 & 0 \end{pmatrix}$$

blocks on the diagonal and 0's elsewhere. The orbit $G \cdot J$ is closed and, in fact, is $\{s \in S_{2n} : \mathrm{Pf}(s) = 1\}$ where Pf is the Pfaffian. Let H be the symplectic group, i.e., $H = \{g \in G : gJg^T = J\}$. Define polynomials $D_i \in k[S_{2n}]$, $1 \leq i \leq n$, as follows. Let $s \in S_{2n}$ and let s' be the skew-symmetric matrix obtained from s by deleting its first $2i$-2 rows and $2i$-2 columns; then $D_i(s) = \mathrm{Pf}(s')$. Then, D_i is U-invariant, has T-weight ω_{2i-2}, and is irreducible.

Theorem 11.10. *$k[S_{2n}]^U = k[D_1, \ldots, D_n]$, a polynomial algebra.*

Theorem 11.11. *Let H and G be as above. Then H is a spherical, Grosshans subgroup of G. The algebra of U-invariants on G/H is $k[D_2, \ldots, D_n]$, a polynomial algebra over k. If char $k = 0$ and if Z is any rational G-module, then the mapping $(C[S_n] \otimes Z)^G$ to Z^H given by $\Sigma_i f_i \otimes z_i \rightarrow \Sigma_i f_i(J) z_i$ is surjective.*

D. Euclidean geometry

> *This whole manner of viewing the subject [the structure of geometry] was given an important turn by the great English geometer A. Cayley in 1859. Whereas, up to this time, it had seemed that affine and projective geometry were poorer sections of metric geometry, Cayley made it possible, on the contrary, to look upon affine geometry as well as metric geometry as special cases of projective geometry; "projective geometry is all geometry." This apparently paradoxical connection arises from the fact the one adjoins to the figures under investigation certain manifolds, namely, the plane at infinity or as the case may be, the imaginary spherical circle which lies on it; hence the affine or the metric properties, respectively, of a figure are nothing but the projective properties of the figure thus extended.* [59; p.134]

Let $G = SL(\mathrm{n},\mathbf{C})$ and let H be the subgroup of G characteristic of metric geometry, namely, the set of all matrices having the form

$$\begin{pmatrix} A & b \\ 0 & 1 \end{pmatrix}$$

where $A \in SL(n\text{-}1,\mathbf{C})$, $A^TA = I$, and b is an $(n\text{-}1) \times 1$ column matrix in $\mathbf{C}^{n-1}$. Let $V = \{(a_1, \ldots, a_n) : a_i \in \mathbf{C}\}$. The group G acts on V by $g{\cdot}r = rg^{-1}$ for all $g \in G$ and $r \in V$. Let S_n be the set of all $n \times n$ symmetric matrices with entries in $\mathbf{C}$. The group acts on S_n by $g{\cdot}s = gsg^T$ for all $g \in G$ and $s \in S_n$. Then G acts on the vector space $V \times S_n$ by $g{\cdot}(r,s) = (rg^{-1}, gsg^T)$. Let

$$v = (0, \ \ldots \ ,0,1) \times \begin{pmatrix} I_{n-1} & 0 \\ 0 & 0 \end{pmatrix}.$$

A straightforward calculation shows that $H = G_v$.

Theorem 11.12. *Let G, H, and v be as above. Let $X = cl(G{\cdot}v)$. Then H is a spherical, Grosshans subgroup of G and $C[X] = C[G/H]$. The U-invariants on X constitute a polynomial algebra generated by x_1 and D_i, $3 \leq i \leq n$, where the D_i are as in Theorem 11.8 and $x_1(a_1, \ldots , a_n) = a_1$. If Z is any rational G-module, then the mapping $(C[V \times S_n] \otimes Z)^G$ to Z^H given by $\Sigma_i f_i \otimes z_i \rightarrow \Sigma_i f_i(v)z_i$ is surjective.*

Proof. Let B^- be the Borel subgroup of G consisting of all lower triangular matrices in G. Then $G_v \cap B^-$ is finite and a dimension calculation shows that $B^-{\cdot}v$ is open in X so H is spherical. The fact about the U-invariants is left as Exercise 5. That $\mathbf{C}[G/H]$ is finitely generated over $\mathbf{C}$ will follow immediately once we show that $\mathbf{C}[X] = \mathbf{C}[G/H]$. To that end, let M_n be the set of all $n \times n$ matrices over $\mathbf{C}$. The group G acts on M_n by right translation, namely, $g{\cdot}m = mg^{-1}$ for all $g \in G$ and $m \in M_n$. We may also consider M_n to be $V \times \ldots \times V$ (n-times), i.e., a product of its rows. Let $x = (a_1, \ldots , a_n) \in V$. We put $x' = (a_1, \ldots, a_{n-1})$ and write elements $(x_1, \ldots, x_n) \in V \times \ldots \times V$ as

$$\begin{matrix} x_1{}' & a_n \\ x_2{}' & b_n \\ & \vdots \\ x_n{}' & c_n. \end{matrix}$$

The action of H on $V \times \ldots \times V$ (n-times) is that of a group of "step transformations"; a theorem of Weyl [116; Theorem (2.7A), p.47] along with the "first fundamental theorem" for orthogonal groups [116; Theorem (2.9A), p.53] then gives the generators for the algebra of invariants as follows: (i) $x_i{}'x_j{}'^T$ for $1 \leq i \leq j \leq n$; (ii)

$$[x'_{i_1} \ \ldots \ x'_{i_{n-1}}]$$

where $i_1, \ldots, i_{n-1}$ are $(n\text{-}1)$ distinct indices chosen from $\{1,2,\ldots,n\}$ and $[xy\ldots z] = \det(x,y, \ldots,z)$; (iii) $[x_1, \ldots, x_n] = \det(x_1, \ldots, x_n)$.

Now, consider the restriction mapping from $\mathbf{C}[M_n] = \mathbf{C}[V \times \ldots \times V]$ to $\mathbf{C}[SL_n]$.

Since actions of SL_n are completely reducible, $\mathbf{C}[V \times \ldots \times V]^H$ is sent onto $\mathbf{C}[SL_n]^H$ (Exercise 6). Thus, $\mathbf{C}[SL_n]^H$ is generated by the restrictions of the invariants having forms (i) and (ii) above.

Let φ: $\mathbf{C}[X] \to \mathbf{C}[SL_n]^H$ be the mapping defined by $(\varphi f)(g) = f(g \cdot v)$. Let $s_{i,j}$ be the (i,j) co-ordinate function on S_n and x_i be the ith co-ordinate function on V. Then $\varphi(s_{i,j})$ gives all the generators of type (i) and $\varphi(x_i)$ of type (ii) (using the descriptions of g^{-1} in terms of the adjoint matrix of g). Thus, $\mathbf{C}[X] = \mathbf{C}[G]^H$. From Lemmas 1.8 and 4.2, it follows that X is normal and that each irreducible component of $X - G\cdot v$ has codimension ≥ 2 in X. We then may apply Lemma 11.5 to relate the invariants of G to those of H. QED

Note. The approach to Euclidean geometry just outlined may be used effectively in "automatic proving of geometric theorems" [20].

E. Hilbert's example

> *Let us mention another special invariant theory. Suppose we are given a space curve of order three. One can ask about those transformations that leave this space curve unchanged. These, of course, form a group, namely - if we focus on a particular example (cf Hilbert 1890, p.533) - a four parameter group of quaternary substitutions. In this case too, there is a process Ω, analogous to the previous one, so that everything procedes as before. At the same time, there exists a certain quaternary form F of order four that remains unchanged under these substitutions. This form is the discriminant of the cubic equation in t: $x_1t^3 + x_2t^2 + x_3t + x_4 = 0$. Here, too, the question arises whether the invariants consist of the general invariants of f and those of f and F.* [46, p.186]

We begin by putting Hilbert's question into modern terms. Let $G = SL_2(\mathbf{C})$, let T be the group of diagonal matrices in G, and let ρ be the four-dimensional representation of G on V_3 as defined in §10. Let $\{x^3, 3x^2y, 3xy^2, y^3\}$ be a basis for V_3; if $v \in V_3$, we shall write $v = ax^3 + 3bx^2y + 3cxy^2 + dy^3$. With respect to this basis, we shall consider ρ to be a homomorphism from $SL_2(\mathbf{C})$ into $SL_4(\mathbf{C})$. There is an invariant $D \in \mathbf{C}[V_3]^G$, namely, $D = a^2d^2 - 3b^2c^2 + 4b^3d + 4ac^3 - 6abcd$ (see §10). Now, in the language of these lecture notes, Hilbert is asking this. If A is any rational $SL_4(\mathbf{C})$ - algebra, is A^G equal to the algebra of invariants of $SL_4(\mathbf{C})$ on $(\mathbf{C}[SL_4(\mathbf{C})\cdot D] \otimes A)$? We are going to see that the answer is negative in general.

Let $H = \{g \in SL_4(\mathbf{C}) : g\cdot D = D\}$. A calculation using Lie algebras shows that $\rho(G) = H^0$, the connected component of the identity element in H. Let N be the normalizer of $\rho(G)$ in $SL_4(\mathbf{C})$. We next show that $N = H$ and the index of $\rho(G)$ in N is 2. To do this, let n be an element in N which is not in $\rho(G)$. We may suppose that n normalizes T. If $t \in T$ has diagonal entries α, α^{-1}, then $\rho(t)$ is the diagonal matrix with entries $\alpha^{-3}, \alpha^{-1}, \alpha$, and α^3. Thus, the distinct weight spaces of T on V_3 are all of dimension 1. The element n must permute these four subspaces. Also, since $n\cdot D$ is in

$\mathbf{C}[V_3]^G$, $n{\cdot}D$ must be a scalar multiple of D. In this way, we see that n must be of the following two forms:

$$\begin{pmatrix} \alpha & 0 & 0 & 0 \\ 0 & \beta & 0 & 0 \\ 0 & 0 & \gamma & 0 \\ 0 & 0 & 0 & \delta \end{pmatrix}, \begin{pmatrix} 0 & 0 & 0 & \alpha \\ 0 & 0 & \beta & 0 \\ 0 & \gamma & 0 & 0 \\ \delta & 0 & 0 & 0 \end{pmatrix}.$$

Since $n{\cdot}D$ is a scalar multiple of D, we see that $\alpha^2\delta^2 = \beta^2\gamma^2 = \alpha^3\gamma = \beta^3\delta = \alpha\beta\gamma\delta = 1$. Hence, $\alpha\delta = \beta\gamma$, $\gamma^2 = \beta\delta$, $\beta^2 = \alpha\gamma$. If we multiply the second and third equalities and use the fact that $\alpha\beta\gamma\delta = 1$, we see that $\beta = \pm\gamma^{-1}$, $\alpha = \gamma^{-3}$, $\delta = \pm\gamma^3$. Thus, modifying each matrix by a suitable element in $\rho(T)$, we may assume that n has one of the following forms :

$$\begin{pmatrix} 1 & 0 & 0 & 0 \\ 0 & -1 & 0 & 0 \\ 0 & 0 & 1 & 0 \\ 0 & 0 & 0 & -1 \end{pmatrix}, \begin{pmatrix} 0 & 0 & 0 & 1 \\ 0 & 0 & 1 & 0 \\ 0 & 1 & 0 & 0 \\ 1 & 0 & 0 & 0 \end{pmatrix}.$$

The product of these two matrices is in $\rho(G)$ since it is the image of

$$\begin{pmatrix} 0 & 1 \\ -1 & 0 \end{pmatrix}.$$

Alternatively, we could take n to be the diagonal matrix in $SL_4(\mathbf{C})$ with entries, i,i,i,i on the diagonal where $i^2 = -1$. It will be important later to note that this n is the image in $SL_4(\mathbf{C})$ of the diagonal matrix n' with diagonal entries $-i,-i$ in $GL_2(\mathbf{C})$. Thus, we have shown that $N = H = \rho(G) \cup n\rho(G)$. The homogeneous space $SL_4(\mathbf{C})/H$ is isomorphic to $SL_4(\mathbf{C}){\cdot}D$. In fact, we see that the orbit $SL_4(\mathbf{C}){\cdot}D$ is closed using the following result of D. Luna.

Theorem 11.13 [69; Corollaire 1, p. 237]. *Let G be a connected reductive group and let H be a reductive subgroup of G. Let X be an affine G - variety and let $x \in X$ be fixed by H. Then the orbit $G{\cdot}x$ is closed if and only if $N_G(H){\cdot}x$ is closed.*

We should note that since the orbit $SL_4(\mathbf{C}){\cdot}D$ is closed, the algebra of invariants of $SL_4(\mathbf{C})$ on $(\mathbf{C}[SL_4(\mathbf{C}){\cdot}D] \otimes A)$ is actually the homomorphic image under the restriction mapping of the algebra of invariants of $SL_4(\mathbf{C})$ on $(\mathbf{C}[V_3] \otimes A)$. Now, let A be any rational $SL_4(\mathbf{C})$-algebra. According to Theorem 9.1, the algebra of invariants of $SL_4(\mathbf{C})$ on $(\mathbf{C}[SL_4(\mathbf{C}){\cdot}D] \otimes A) = (\mathbf{C}[SL_4(\mathbf{C})/H] \otimes A)$ is A^H. Let us take A to be $\mathbf{C}[SL_4(\mathbf{C})]$ where $SL_4(\mathbf{C})$ acts by right translation. Both the subgroups $\rho(G)$ and H of $SL_4(\mathbf{C})$ are observable since they are reductive (Corollary 2.4), so the algebra A^H is properly contained in A^G. Thus, the answer to Hilbert's question is negative in general.

One final point may be in order. For Hilbert the algebras A of interest are $\mathbf{C}[\mathcal{S}^r(V_3)]$, the algebras of regular functions on symmetric powers of V_3. Even here, the answer is negative. For example, let us consider $S^3(V_3)$. This is a $GL_2(\mathbf{C})$-module under the plethysm $GL_2(\mathbf{C}) \to GL_4(\mathbf{C}) \to GL(S^3(V_3))$. In the usual notation, as a $GL_2(\mathbf{C})$ -

module, this module decomposes as 6,3 ⊕ 7,2 ⊕ 9. The image of n' is the diagonal matrix with $-i$ on the diagonal. An invariant of G on $S^3(V_3)$ is the discriminant of the quadratic form corresponding to the representation 6,3. But this discriminant is not invariant under n'.

Note. An algorithm for plethysms arising from the embedding $GL_2(\mathbf{C}) \to GL_4(\mathbf{C})$ is given in [81].

Exercises.

1. Let G be connected, reductive. Prove that the subgroup $G_1 = \{(g,g) : g \in G\}$ is spherical in $G \times G$.
2. Let A be a commutative k-algebra which is a unique factorization domain and which has k^* as its group of units. Let H be a connected, solvable linear algebraic group which acts rationally on A. Let $a \in A$ be a H-weight function and let $a = p_1 \ldots p_r$ be the decomposition of a into irreducible factors. Show that each p_i is a H-weight function. Let $U = \mathfrak{R}_u H$. Show, also, that A^U is a unique factorization domain. (Hint: let $h \in H$. Then $h \cdot a = \chi(h)a$ so $(h \cdot p_1) \ldots (h \cdot p_r) = \chi(h)p_1 \ldots p_r$. Thus, $h \cdot p_i = cp_i$ for all i.)
3. Prove the sufficiency in each of the statements in Lemma 11.5.
4. Let char $k = 0$. Let G be a connected, reductive group with Borel subgroup $B = TU$, let $\omega \in X(T)$ and let $E(\iota\omega) = \{f \in k[G]^U : r_t f = \omega(t)f \text{ for all } t \in T\}$. Prove that $E(\iota\omega)$ is irreducible under the action of G by left translation. (Hint: First, $E(\iota\omega)$ has finite dimension by Exercise 3, §5. Next, let $B^- = TU^-$ be the Borel subgroup "opposite" to B. Let $W \subset E(\iota\omega)$ be the set of elements fixed under left translation by U^-. Then dim $W = 1$ since U^-TU is open in G. Thus, the subspace of $E(\iota\omega)$ consisting of elements fixed by U has dimension = 1.)
5. Prove the statement about U-invariants in Theorem 11.12. (Hint: (i) The algebra $\mathbf{C}[G]$ is known to be a unique factorization domain [86; Corollary, p.303]. (ii) Show that the character group of H is trivial. (iii) Show that $\mathbf{C}[G/H]$ is a unique factorization domain by an argument similar to that given in Exercise 2. (iv) Consider the subvariety S' of X consisting of all elements of the form $(x, 0, \ldots ,0) \times s$, where $s = (s_{ij})$ is a diagonal matrix such that $s_{11} = 0$ and $s_{22} \ldots s_{nn} = 1$. Using S', show that $k(X)^U = k(x_1, D_3, \ldots , D_n)$. (v) Show that x_1 has T-weight ω_1 and that D_i has T-weight $2\omega_{i-1}$. (vi) Show that x_1 and the D_i are irreducible by considering T-weights. (vii) Conclude that the statement about U-invariants holds.)
6. Let char $k = 0$, let G be a reductive group, and let H be a subgroup of G. Let V_1, V_2, and V_3 be G-modules along with G-linear transformations such that the sequence $0 \to V_1 \to V_2 \to V_3 \to 0$ is exact. Show that the sequence $0 \to V_1^H \to V_2^H \to V_3^H \to 0$ is exact.
7 [53, p.315]. Let char $k = 0$. Let V be a vector space over k having dimension n. Let $\Lambda(V)$ be the exterior algebra over V and let $\Lambda^r(V)$ be the subspace of $\Lambda(V)$ spanned by all elements having degree r. Let Z_r be the Zariski closure in $\Lambda^r(V)$ of $\{v_1 \wedge v_2 \wedge \ldots \wedge v_r : v_i \in V\}$. Let $G = GL(V)$ and let $B = TU$ be the Borel subgroup of G consisting of all upper triangular matrices in G relative to some basis $\{e_1, \ldots , e_n\}$ of V. Suppose that $r \geq s$ and $r + s \leq n$. Let $T_1 = \{(a,b) : ab \neq 0\}$ and let T_1 act on $Z_r \times Z_s$ by $(a, b)\cdot(z, z') = (az, bz')$. Then T_1B has an open orbit on $Z_r \times Z_s$. In fact, show that the T_1B-orbit of (z_1, z_2) is open where $z_1 = e_n \wedge e_{n-1} \wedge \ldots \wedge e_{n-r+1}$ and $z_2 = (e_n + e_{n-r-s+1}) \wedge (e_{n-1} + e_{n-r-s+2}) \wedge \ldots \wedge (e_{n-s+1} + e_{n-r})$. (The algebra of invariants $k[Z_r \times Z_s]^U$ may

be described explicitly. Then, $k[Z_r \times Z_s \times Z_p \times Z_q]^G$ may be calculated following the procedure outlined above. Remarkably enough, it is always a polynomial algebra.)

8. Show that the determinant of a symmetric matrix (s_{ij}) is irreducible. (Hint: modify the usual proof. Namely, suppose that the determinant factors, say, PQ. If s_{11} appears in P, then only terms in its cofactor can appear in Q. Now, let $r > 1$ and look at s_{rr}. If s_{rr} appears in Q, then s_{r1} cannot appear in either P or Q. Thus, all the s_{rr} appear in P.)

9. Prove Theorem 11.11. (Hint: show that under the action of U, a skew-symmetric matrix can be changed to a matrix which is the sum of blocks having the form

$$\begin{pmatrix} 0 & x \\ -x & 0 \end{pmatrix}.)$$

Chapter Three

Invariants of Maximal Unipotent Subgroups

Introduction. Let G be a reductive group and let U be a maximal unipotent subgroup of G. Let A be a commutative k-algebra on which G acts rationally. In this chapter, we compare properties of A and A^U. A key tool is the description of the representations $E(\omega) \subset k[G]^U$ given in §12; this is illustrated in §13 with the example $G = GL_n$. Then we study the relationship between the algebras A and $G{\cdot}A^U$ in §14, proving that A is integral over $G{\cdot}A^U$. In §15, we construct (following ideas in the case char $k = 0$ found by D. Luna) a graded algebra, gr$A$, which then becomes a kind of machine allowing results in characteristic 0 to be proved for arbitrary characteristic. First, using this algebra, we show in §16 that A is finitely generated over k if and only if A^U is; we also show that an A-module M is finitely generated if and only if the A^U-module M^U is finitely generated. Next, observable subgroups which contain a maximal unipotent subgroup have particularly good properties; these are studied in §17. In §18, other properties of A and A^U are related including the presence of nilpotents, zero-divisors, and normality. Throughout this chapter, we follow the notation and terminology for the group G° given at the beginning of §3 (unless otherwise noted)

§12. The Representations $E(\omega)$

- In this section, we shall assume that G is connected.

Let H be a subgroup of G. The group H acts on $k[G]$ by left translation (denoted by ℓ_g or $\ell(g)$) and right translation (denoted by r_g or $r(g)$). If V is a subspace of $k[G]$ which is invariant under right (resp. left) translation by H, we put V^H (resp. ${}^HV$) $= \{v \in V : r_h v = v \text{ for all } h \in H\}$ (resp: $\{v \in V : \ell_h v = v \text{ for all } h \in H\}$). In particular, since T normalizes U, T acts (rationally) on $k[G]^U$ by right translation.

Definition. Let $\omega \in X(T)$. Then $E(\omega) = \{f \in k[G]^U : r(t)f = (\iota\omega)(t)f \text{ for all } t \in T\}$.

Each $E(\omega)$ is an induced module (Exercise 1) and is $\{0\}$ unless ω is dominant (Theorem 3.2). (The $E(\omega)$ are the contragredients of the "Weyl modules" as defined in [57].) Since each element in $k[G]^U$ is a direct sum of T-weight vectors (where T acts by right translation), we see that $k[G]^U$ is a direct sum of the $E(\omega)$, $\omega \in X^+(T)$. From the definition, it is obvious that if $\omega, \omega' \in X^+(T)$, then $E(\omega)E(\omega') \subset E(\omega + \omega')$.

Theorem 12.1. *Let* $\omega \in X^+(T)$.
(a) *The subspace* $E(\omega)$ *is non-zero.*
(b) *The group* G *acts on* $E(\omega)$ *by left translation and* $\dim({}^UE(\omega)) = 1$. *If* $f \in {}^UE(\omega)$, *then* $\ell(t)f = \omega(t)f$ *for all* $t \in T$.
(c) *The subspace* $E(\omega)$ *is a finite-dimensional vector space over* k.

(d) *If char k = 0, the space $E(\omega)$ is the irreducible representation of G having highest weight ω. If char k = p $\geq$ 0, the space $E(\omega)$ contains the irreducible representation of G having highest weight ω.*

Proof. Let V be the irreducible, rational G-module having $\iota\omega$ as highest weight and let v be a highest weight vector with respect to B in V. Let V^* be the dual space of V and define a mapping $\varphi: V^* \to k[G]$ by $(\varphi\mu)(g) = \mu(g{\cdot}v)$ for all $\mu \in V^*$, $g \in G$. Obviously, the image of V^* under φ is a non-zero subspace of $k[G]$. Furthermore, for $b = tu \in TU$ we have $(r(b)\varphi\mu)(g) = (\varphi\mu)(gb) = \mu(gb{\cdot}v) = \mu(g(\iota\omega)(t){\cdot}v) = (\iota\omega)(t)\mu(g{\cdot}v) = (\iota\omega)(t)(\varphi\mu)(g)$. Thus, $\varphi\mu \in E(\omega)$ and (a) is proved.

The group G acts on $E(\omega)$ by left translation since $\ell(g)r(g') = r(g')\ell(g)$ for all g, $g' \in G$. Any element v in $E(\omega)$ is contained in a finite-dimensional subspace which is invariant with respect to left translation by G. (In fact, the set of all $\ell(g){\cdot}v$, $g \in G$, spans such a space.) In any such non-zero subspace, we may find non-zero elements fixed by U^-. Let f be such an element, i.e., $f \neq 0$, $\ell(u_1)f = f$, $r(t)f = (\iota\omega)(t)f$, and $r(u)f = f$ for all $u \in U$, $t \in T$, and $u_1 \in U^-$. Then, for any $u_1 \in U^-$, $t \in T$, and $u \in U$ we have $f(u_1tu) = f(u_1etu) = (\ell(u_1^{-1})r(tu)f)(e) = (\iota\omega)(t)f(e)$. From this and the fact that the big open cell U^-TU is dense in G, we may conclude the following: any two non-zero functions in $E(\omega)$ fixed by U^- differ by at most a constant, i.e., the dimension of U^--fixed points in $E(\omega)$ is 1.

Let f be as in the preceding paragraph; we next show that $\ell(t)f = (s_o\omega)(t)f$ for all $t \in T$. Let $u_1 \in U^-$, $t_1 \in T$, and $u \in U$. Then $(\ell(t)f)(u_1t_1u) = f(t^{-1}u_1t_1u) = f(t^{-1}u_1tt^{-1}t_1u)$. Since $t^{-1}u_1t \in U^-$, we see as before that this is $(\iota\omega)(t^{-1}t_1)f(e) = (s_o\omega)(t){\cdot}(\iota\omega)(t_1)f(e) = (s_o\omega)(t)f(u_1t_1u)$. Since $s_o^{-1}Us_o = U^-$, we see that $\ell(s_o)$ sends ${}^{U^-}E(\omega)$ onto ${}^{U}E(\omega)$ and that $\ell(s_o)f$ has T-weight $s_o{\cdot}(s_o{\cdot}\omega) = \omega$. This completes the proof of (b).

Now, we prove statement (c) (following an argument given in [29; (1.5.2), p.17]); we show first that there are only finitely many weights of T acting on $E(\omega)$ by left translation and, then, that each such weight space is finite-dimensional. Let $\chi \in X(T)$ and let $f \in E(\omega)$ be a non-zero element such that $\ell(t)f = \chi(t)f$ for all $t \in T$. Let $\langle U{\cdot}f\rangle$ denote the smallest subspace of $E(\omega)$ which is invariant under all $\ell(u)$, $u \in U$; this is the subspace spanned by all $\ell(u)f$, $u \in U$. This subspace is invariant under each $\ell(t)$ since $\ell(t)\ell(u)f = \ell(tut^{-1})\ell(t)f = \chi(t)\ell(tut^{-1})f$. Furthermore, if χ' is any weight of T on $\langle U{\cdot}f\rangle$, then $\chi' \geq \chi$ (by Lemma 3.1). There is a non-zero element v in $\langle U{\cdot}f\rangle$ which is fixed by U (under left translation) and v has T-weight ω by statement (b). Thus, $\chi \leq \omega$. We may apply this same argument using U^- instead of U to see that $s_o\omega \leq \chi$. It follows that there are only finitely many weights of T acting on $E(\omega)$ by left translation and that any such weight χ must satisfy the inequalities $s_o\omega \leq \chi \leq \omega$.

To complete the proof of statement (c), it is enough to show that each weight space of T acting on $E(\omega)$ by left translation is finite-dimensional. As a variety, the group B^- is isomorphic to the product $T \times \Pi\, U_\alpha$ where α ranges over all the negative roots in $\Phi(T,G)$. Let x_β be the co-ordinate function on $T \times \Pi\, U_\alpha$ corresponding to the root β. A direct computation shows that $r(t)x_\beta = (-\beta)(t)x_\beta$ and $\ell(t)x_\beta = x_\beta$ for all $t \in T$. (For example, $(r(t)x_\beta)(t_1\Pi u_\alpha(\xi_\alpha)) = x_\beta(t_1\Pi u_\alpha(\xi_\alpha)t) = x_\beta(t_1t\Pi t^{-1}u_\alpha(\xi_\alpha)t) = x_\beta(u_\beta(-\beta(t)\xi_\beta)) = -\beta(t)\xi_\beta = -\beta(t)x_\beta(t_1\Pi u_\alpha(\xi_\alpha))$. Furthermore, if $\chi \in X(T)$ is considered to be a function on B^-, then $r(t)\chi = \chi(t)\chi$ and $\ell(t)\chi = (-\chi)(t)\chi$ for all $t \in T$. Let

$$m = \chi' \prod x_\alpha^{e_\alpha}$$

be a function on B^- where $\chi' \in X(T)$ and the e_α are non-negative integers. Then m has weight $\chi' - \Sigma e_\alpha \alpha$ (resp. $-\chi'$) when T acts by right translation (resp. left translation).

Let π: $k[G] \rightarrow k[B^-]$ be the restriction mapping. Then π is injective on $E(\omega)$. Indeed, let $f \in E(\omega)$ and suppose that $\pi(f) = 0$. Then for any $b \in B^-$ and $u \in U$, we have $f(b^-u) = (r(u)f)(b^-) = f(b^-) = 0$. Since $f = 0$ on $B^- \times U$, we see that $f = 0$.

Finally, let $f \in E(\omega)$ and suppose that $\ell(t)f = \chi(t)f$ for all $t \in T$. Then $\pi(f)$ is a linear combination of monomials m each having the form

$$m = -\chi \prod x_\alpha^{e_\alpha}$$

where $-\chi - \Sigma e_\alpha \alpha = \iota\omega$. Since $s_o\omega \leq \chi \leq \omega$, we see that $s_o\omega - \omega \leq \Sigma e_\alpha \alpha \leq 0$. The space spanned by all such m is (obviously) finite-dimensional and statement (c) is proved.

To prove (d), we use the notation in the first paragraph of this proof. Then, since V^* is the irreducible representation of G having highest weight ω, V^* must be isomorphic to a subspace of $E(\omega)$ since the mapping φ is G-equivariant (Exercise 3, §1). In case char $k = 0$, the actions of G are completely reducible. That is, any finite-dimensional, rational G-module is a direct sum of irreducible G-modules, each having a unique highest weight vector with respect to B. However, dim ${}^UE(\omega) = 1$ according to (b). QED

Let A be a commutative k-algebra on which G acts rationally. In Section 16, we shall show that A is finitely generated over k if and only if A^U is. We are particularly interested in the case where $A = k[G]^H$, H being a closed subgroup of G, and G acting by left translation. By the result to be proved in Section 16, $k[G]^H$ is finitely generated over k if and only if ${}^Uk[G]^H$ is. In the study of ${}^Uk[G]^H$, one is naturally led to T-weight spaces; for $\omega \in X^+(T)$, we wish to consider $\{f \in k[G]^H : \ell_{tu}f = \omega(t)f$ for all $t \in T$, $u \in U\}$. We shall show that the dimension of this space is dim ${}^HE(\iota\omega)$, a result known (in the case char $k = 0$) as "Frobenius reciprocity".

Theorem 12.2. *Let H be a closed subgroup of G. Let $\omega \in X^+(T)$. Then dim ${}^HE(\iota\omega) =$ dim $\{f \in k[G]^H : \ell(b)f = \omega(b)f$ for all $b \in B\}$, the multiplicity of ω in $k[G]^H$.*

Proof. The mapping ϵ: $k[G] \rightarrow k[G]$ given by $(\epsilon f)(x) = f(x^{-1})$ satisfies the condition that $r_g \circ \epsilon = \epsilon \circ \ell_g$ for all $g \in G$. The theorem follows immediately. QED

Corollary 12.3. *Let H be a spherical subgroup of G. Then dim${}^HE(\omega) \leq 1$ for all $\omega \in X^+(T)$.*

Proof. According to the definition given in §11, H is spherical if the action of G by left translation on $k[G/H]$ is multiplicity-free, i.e., for each $\omega \in X(B)$, the subspace $\{z \in k[G]^H : \ell(b)z = \omega(b)z$ for all $b \in B\}$ is of dimension ≤ 1. Thus, Corollary 12.3 follows immediately from Theorem 12.2. QED

According to J. A. Green [36, p.33], the next result can be traced back to J. Deruyts and his seminal work on the representations of GL_n, "Essai d'une théorie générale des formes algébriques" (Mém. Soc. Roy. Liège **17** (1892), pp 1 - 156). We shall encounter it in

§16 and §17.

Theorem 12.4. *Let $\omega \in X^+(T)$.*
(a) *There is a vector $v \in E(\omega)^*$ such that $\ell(tu)\cdot v = (\iota\omega)(t)v$ for all $t \in T$, $u \in U$, and $E(\omega)^* = \langle G\cdot v\rangle$.*
(b) *Let Z be any rational G-module and let $(k[G] \otimes Z)^G = \{\Sigma f_i \otimes z_i \in k[G] \otimes Z : \Sigma \ell_g f_i \otimes g\cdot z_i = \Sigma f_i \otimes z_i$ for all $g \in G\}$. Let Φ: $(k[G] \otimes Z)^G \to Z$ be defined by $\Phi(\Sigma f_i \otimes z_i) = \Sigma f_i(e)z_i$.*

(i) *Let $Z(U,\omega) = \{z \in Z : tu\cdot z = \omega(t)z$ for all $t \in T$ and $u \in U\}$. Then Φ defines an isomorphism from $(E(\iota\omega) \otimes Z)^G$ onto $Z(U,\omega)$.*

(ii) *Let $z \in Z(U,\omega)$ and suppose that $\Phi(\Sigma f_i \otimes z_i) = z$, where $\{f_1, \ldots ,f_m\}$ is a basis of $E(\iota\omega)$ and $\Sigma f_i \otimes z_i$ is G-invariant. Let $\{f_1^*, \ldots , f_m^*\}$ be the dual basis of $E(\iota\omega)^*$. Then the mapping ψ: $E(\iota\omega)^* \to Z$ given by $\psi(f_i^*) = z_i$ is G-equivariant and the elements $z_1, \ldots ,z_m$ span $\langle G\cdot z\rangle$. If char $k = 0$, then ψ is an isomorphism.*

Proof. The weights of T on $E(\omega)^*$ are of the form $-\chi$ where χ is a weight of T on $E(\omega)$; we also saw in the proof of Theorem 12.1(c) that for such χ, $s_o\omega \leq \chi \leq \omega$. Let v be a weight vector in $E(\omega)^*$ corresponding to $-s_o\omega = \iota\omega$. The vector $s_o v$ has weight $-\omega$; if $w \in E(\omega)$ is a highest weight vector, then $s_o v(w)$ is non-zero. By our choice of v, $\ell_{tu}\cdot v = (\iota\omega)(t)v$ for all $t \in T$, $u \in U$. Now suppose that $\langle G\cdot v\rangle$ is a proper subspace of $E(\omega)^*$. Then, we may find a non-zero $z \in E(\omega)$ such that $f(z) = 0$ for all $f \in \langle G\cdot v\rangle$. Since the space $\langle G\cdot v\rangle$ is G-invariant, we have $v(g\cdot z) = 0$ for all $g \in G$. Now, the subspace $\langle G\cdot z\rangle$ is also G-invariant and so contains a non-zero U-fixed point which, by Theorem 12.1(b), we may assume to be w. But as we have seen, $v(s_o w)$ is not zero.

Next, we prove (b). We have seen (Theorem 9.1b) that the mapping Φ is a B-equivariant vector space isomorphism, where B acts on $k[G]$ (and, hence, on $(k[G] \otimes Z)^G$) by right translation. Statement (i) follows immediately.

To prove statement (ii), suppose that $\ell_g f_j = \Sigma t_{ij}(g)f_i$. Since $\Sigma f_j \otimes z_j$ is G-invariant, a straightforward computation shows that $g\cdot z_i = \Sigma t_{ij}(g^{-1})z_j$. Thus, we see that ψ is G-equivariant; also, $z = \Sigma_i f_i(e)z_i$, is (obviously) a linear combination of the z_i. According to (a), $E(\iota\omega)^* = \langle G\cdot f\rangle$. The big cell U^-TU is open and dense in G so $\langle G\cdot f\rangle = \langle U^-TU\cdot f\rangle = \langle U^-\cdot f\rangle$ since any linear functional vanishing on $g\cdot f$, $g \in U^-TU$, must also vanish for all $g\cdot f$, $g \in G$. It follows from Lemma 3.1 that any weight vector of T on $\langle G\cdot f\rangle$ other than f corresponds to a weight $\chi < \omega$. This shows that $\psi(f) = z$ and, since ψ is G-equivariant, that $\psi(g\cdot f) = g\cdot z$ for all $g \in G$. If char $k = 0$, then both $E(\omega)$ and $E(\omega)^*$ are irreducible, so ψ is injective. QED

- In the rest of this section, we shall assume that char $k = 0$.

Lemma 12.5. *Let V be a rational G-module and let $v \in V$ be a non-zero vector such that $b\cdot v = \omega(b)v$ for all $b \in B$. Then $\langle G\cdot v\rangle$ is an irreducible G-module having highest weight ω.*

Proof. The subspace $\langle G\cdot v\rangle$ is finite-dimensional since V is a rational G-module. Also, as we saw in the proof of Theorem 12.4(b)(ii), $\langle G\cdot v\rangle = \langle U^-\cdot v\rangle$ and any weight vector of T on $\langle G\cdot v\rangle$ other than v corresponds to a weight $\chi < \omega$. Since the actions of G are completely reducible, we may write $\langle G\cdot v\rangle$ as a direct sum of irreducible G-modules $V(\omega_1), \ldots , V(\omega_r)$. There can be one and only one ω_i with $\omega_i =$

ω and then $<G{\cdot}v> \subset V(\omega_i)$. QED

Theorem 12.6. *Let $\omega, \omega' \in X^+(T)$. The mapping $E(\omega) \otimes E(\omega') \to E(\omega + \omega')$ given by $\Sigma a_i \otimes b_i \to \Sigma a_i b_i$ is surjective.*
Proof. The subspace $E(\omega)E(\omega')$ is a non-zero G-invariant subspace of $E(\omega + \omega')$. We apply Theorem 12.1(d) to see that $E(\omega)E(\omega') = E(\omega + \omega')$. QED

Corollary 12.7. *Let G be simply connected and semi-simple. Then, $k[G]^U$ is generated by the $E(\omega_i)$, where the ω_i are the (finitely many) fundamental weights of G.*
Proof. Any highest weight ω has the form $e_1\omega_1 + \ldots + e_r\omega_r$ where the e's are non-negative integers. Then

$$E(\omega) = E(\omega_1)^{e_1} \ldots E(\omega_r)^{e_r}$$

according to Theorem 12.6. QED

It is worth noting that Corollary 12.7 gives a complete representation-theoretic argument for Theorem 5.4 in case char $k = 0$. The key tools are Theorem 12.1, proved in arbitrary characteristic, and Theorem 12.6, proved only when char $k = 0$. However, it is also possible to prove Theorem 12.6 in arbitrary characteristic [57; Proposition, p.413].

Theorem 12.8. *Let Z be any rational G-module. Then, $\oplus_\omega((E(\iota\omega) \otimes Z)^G \otimes E(\omega))$ is isomorphic to Z via the mapping $a \otimes z \otimes b \to <a,b>z$, where the sum is over all dominant weights ω appearing on Z.*
Proof. This is Exercise 4, below.

The group $G \times G$ acts on G by $(a,b){\cdot}x = axb^{-1}$ and this gives an action of $G \times G$ on $k[G]$. (The case char $k > 0$ is discussed in the Note on p.90.)

Theorem 12.9. *As a $G \times G$-module, $k[G]$ is isomorphic to $\oplus_\omega(E(\iota\omega) \otimes E(\omega))$.*
Proof. To keep better track of the G-components, let us denote $G \times G$ by $L \times R$. The $E(\omega)$ for $L \times R$ have the form $E(\eta) \otimes E(\mu)$ where $E(\eta)$ and $E(\mu)$ are the usual induced modules for G. Thus, by Theorem 12.8, $k[G]$ is isomorphic to $\oplus_{\eta,\mu}(E(\iota\eta) \otimes E(\iota\mu) \otimes k[G])^{L \times R} \otimes E(\eta) \otimes E(\mu)$. The actions of L and R commute so this is $\oplus_{\eta,\mu}((E(\iota\eta) \otimes k[G])^L \otimes E(\iota\mu))^R \otimes E(\eta) \otimes E(\mu)$ which, by Theorem 9.1 is $\oplus_{\eta,\mu}(E(\iota\eta) \otimes E(\iota\mu))^R \otimes E(\eta) \otimes E(\mu)$. We now apply Lemma 11.5(b). QED

Exercises.

1. In this exercise, let G be connected (but not necessarily reductive) and let B be a Borel subgroup of G. Let $\lambda \in X(B)$ and let $k(\lambda)$ be the B-module defined by the following action of B on k: $b{\cdot}c = \lambda(b)c$ for all $b \in B$, $c \in k$. (a) For reductive G, show that $E(\omega) = \mathrm{ind}_B{}^G k(s_o\omega)$. (b) In general, show that $\mathrm{ind}_B{}^G k(\lambda)$ is finite-dimensional over k for all λ.
2. We extend Corollary 12.7. Let G be reductive. Show that there exist elements $\omega_1', \ldots, \omega_m'$ in $X^+(T)$ such that $X^+(T)$ is the set of all characters in $X(T)$ having the form

$c_1\omega_1' + \ldots + c_m\omega_m'$ where the c_i's are non-negative integers and that $k[G]^U$ is generated by the (finite-dimensional) subspaces $E(\omega_1'),\ldots,E(\omega_m')$. (Hint: let $f_1,\ldots,f_m$ be elements in $k[G]^U$ and $\omega_1',\ldots,\omega_m'$ be characters in $X^+(T)$ so that (i) $k[G]^U = k[f_1,\ldots f_m]$ and (ii) $r_t f_i = \omega_i'(t)f_i$ for each $i = 1,\ldots,m$. Then $\omega \in X^+(T)$ if and only if $E(\omega) \neq 0$, i.e., if and only if there is an $f \in k[G]^U$ with $f \neq 0$ and $r(t)f = (\iota\omega)(t)f$ for all $t \in T$. But f can be written as a linear combination of terms

$$f_1^{e_1} \cdots f_m^{e_m}$$

each having T-weight $e_1\omega_1' + \ldots + e_m\omega_m'$.)

3. Let H be a subgroup of G. Show that if H fixes non-zero elements in $E(\omega)$ and $E(\omega')$, then H fixes a non-zero element in $E(\omega + \omega')$.

4. Prove Theorem 12.8.

5. For each of the spherical subgroups H of $G = SL(n,k)$ appearing in §11, find all the finite-dimensional irreducible representations of G having an H-fixed point.

6. Let $G = SL(2,k)$. (We follow the notation introduced at the beginning of Section 10.) Show that $E(d\omega) = V_d$.

7. Suppose that G is reductive but not necessarily connected. Let $G = \cup_j g_j G^\circ$ be the coset decomposition of G with respect to G°. For $\omega \in X(T)$, let $E(\omega) = \{f \in k[G]^U : r(t)f = (\iota\omega)(t)f$ for all $t \in T\}$ and $E^\circ(\omega) = \{f \in k[G^\circ]^U : r(t)f = (\iota\omega)(t)f$ for all $t \in T\}$.

(a) Show that $E(\omega) = \oplus_j \ell(g_j)E^\circ(\omega)$.

(b) Show that Theorem 12.4 holds except for the last sentence in (b)(ii).

§13. An Example: The General Linear Group

Certain combinatoric facts about determinants may be used to study the algebras $k[G/H]$ when the group G in question is the general linear group, GL_n. In Part A, we explain the concept of "straightening a bideterminant" and, then, in Part B, show how this may be used to calculate the algebra $k[GL_n/U]$. Straightening tools have been applied by K. Pommerening to the question of the finite generation of $k[GL_n/H]$ for any closed subgroup H which is normalized by a maximal torus. In Part C, we introduce his work.

A. Straightening

Let r be a positive integer. A *partition* λ of r is a sequence of integers $(\lambda_1, \ldots, \lambda_h)$ such that (i) $\lambda_1 + \ldots + \lambda_h = r$ and (ii) $\lambda_1 \geq \lambda_2 \geq \ldots \geq \lambda_h > 0$. If λ and μ are two partitions of r, we shall say that $\mu > \lambda$ if the first non-vanishing difference $\mu_i - \lambda_i$ is positive. To a partition $\lambda = (\lambda_1, \ldots, \lambda_h)$ of r, we associate a *Young diagram*: this consists of a row of λ_1 boxes, then a row of λ_2 boxes below it, and so on with the entire array left justified.

The sequence λ is called the *shape* of the Young diagram. A *Young tableau* is obtained from a Young diagram by placing positive integers in each of the boxes (with one positive integer per box). If T is a Young tableau, the *content* of T is $(\alpha_1, \alpha_2, \ldots)$ where α_i is the number of times the positive integer i appears in T.

If T is any Young tableau, we shall denote by ${}^{C}T$ (resp. ${}^{R}T$) the Young tableau which is obtained from T by writing the entries in each column (resp. each row) of T in non-decreasing order from top to bottom (resp. left to right).

A Young tableau is called *row-injective* if no row has a repeated entry. It is called *normal* if the entries in each row are strictly increasing from left to right. Finally, it is called *standard* if it is normal and the entries in each column are non-decreasing from top to bottom.

Example. Consider the partition (4,3,1) of 8. One possible Young tableau associated to this partition is

$$\begin{matrix} 6 & 4 & 2 & 3 \\ 3 & 1 & 4 & \\ 4. & & & \end{matrix}$$

The content of T is (1,1,2,3,0,1,0, ...). We note that T is not standard since (for example) the entries in the first row are not strictly increasing from left to right. However, the Young tableau ${}^{C}({}^{R}T) = {}^{S}T =$

$$\begin{matrix} 1 & 3 & 4 & 6 \\ 2 & 3 & 4 & \\ 4 & & & \end{matrix}$$

is standard. Quite generally, let T be any row-injective Young tableau. Then, ${}^{C}({}^{R}T) = {}^{S}T$ is always standard and is called the *standardization of T* [22; Lemma 1, p.52].

Now, let m and n be given positive integers. Let x_{ij}, $1 \leq i \leq m$, $1 \leq j \leq n$, be indeterminates over k and let X be the $m \times n$ matrix whose (i,j)-entry is x_{ij}. We shall denote minors of X by $(i_1 \ldots i_p \mid j_1 \ldots j_p)$ where $i_1, \ldots, i_p$ are the row indices and $j_1, \ldots, j_p$ are the column indices.

A *bitableau* $(S \mid T)$ is a pair of Young tableaux S,T having the same shape $\lambda = (\lambda_1, \ldots, \lambda_h)$ where we shall always assume that $\lambda_1 \leq n$. The bitableau is displayed by placing S and T side-by-side. The *content* of $(S \mid T)$ is $(\alpha_1, \alpha_2, \ldots ; \beta_1, \beta_2, \ldots)$ where α_i (resp. β_i) is the number of times that the positive integer i appears in S (resp. T). The bitableau is called *standard* if both S and T are standard Young tableaux.

To a Young bitableau $(S \mid T)$, we may associate a *bideterminant*, i.e., a product of minors of X, in a natural way. For example, to

$$\begin{array}{ccc|ccc} 1 & 3 & 5 & 2 & 4 & 6 \\ 2 & 3 & & 4 & 5 & \\ 4 & & & 5 & & \end{array}$$

we assign the following product of minors: $(1\ 3\ 5 \mid 2\ 4\ 6)(2\ 3 \mid 4\ 5)(4 \mid 5)$.

Theorem 13.1 (the standard basis theorem). *Let* $R = k[x_{11}, \ldots, x_{1n}, x_{21}, \ldots, x_{mn}]$. *The bideterminants of standard bitableaux form a basis over k of R.*

A proof of the standard basis theorem may be found in [23]. We note that "straightening" a bideterminant means expressing the bideterminant in terms of the standard basis. There are constructive, but clumsy, ways to do this one of which comes simply from the Laplace expansion of a determinant.

B. *U*-invariants

Let $GL_n = GL(n,k)$, the group of all $n \times n$ invertible matrices with entries in k. Let U be the standard maximal unipotent subgroup of GL_n, i.e, the subgroup consisting of all upper triangular matrices with 1's on the diagonal. For $1 \leq r < s \leq n$, we shall denote by $U_{r,s}$ the subgroup of U such that $u_{ij} = 0$ for all (i,j) such that $1 \leq i < j \leq n$ unless $i = r$ and $j = s$. Let T be the subgroup of GL_n consisting of all invertible diagonal matrices. Then $B = TU$ is the standard Borel subgroup of GL_n. For $1 \leq r \leq n$, we shall denote the character of T, $t = (t_{ij}) \rightarrow t_{11} \ldots t_{rr}$ by $\omega_r(t)$.

Let $M_{m,n}$ be the space of all $m \times n$ matrices with entries in k. Then $k[M_{m,n}] = k[x_{11},\ldots,x_{1n},x_{21},\ldots,x_{mn}]$ where x_{ij} is the (i,j) co-ordinate function on $M_{m,n}$, i.e., $x_{ij}(m) = m_{ij}$. The group GL_n acts on $M_{m,n}$ by right translation,i.e., $g{\cdot}m = mg^{-1}$ for all $g \in GL_n$, $m \in M_{m,n}$. This action gives rise in the usual way to an action on $k[M_{m,n}]$, namely, $(g{\cdot}f)(m) = f(mg)$. In particular, $g{\cdot}x_{ij} = \Sigma_r x_{ir}g_{rj}$.

Let s be any permutation of $\{1,2,\ldots,n\}$. We shall denote the corresponding permutation matrix by s, also. (This is the matrix with 1's in the $(s{\cdot}i,i)$ entries and 0's elsewhere.) A direct calculation shows that $r_s{\cdot}(i_1 \ldots i_p \mid j_1 \ldots j_p) = (i_1 \ldots i_p \mid sj_1 \ldots sj_p)$.

We wish to give an explicit description of $k[M_{m,n}]^U$. The main tool which we shall employ is, of course, the standard basis theorem.

Lemma 13.2. *Let* $b = tu \in B$ *and let* $f = (i_1 \ldots i_p \mid 1 \ldots p)$. *Then* $b{\cdot}f = \omega_p(t)f$.
Proof. That T acts as described is an immediate consequence of the definitions. To show that U acts as indicated, it suffices to prove that f is fixed by all the $U_{r,s}$. Let $u = u_b \in U_{r,s}$ where $u_{rs} = b$. Then $u{\cdot}x_{ij} = x_{ij}$ if $j \neq s$ and $u{\cdot}x_{is} = bx_{ir} + x_{is}$. Hence, $u{\cdot}f = u{\cdot}(i_1 \ldots i_p \mid 1 \ldots r \ldots s \ldots p) = b(i_1 \ldots i_p \mid 1 \ldots r \ldots r \ldots p) + f = f$. QED

Theorem 13.3. *A basis of* $k[M_{m,n}]^U$ *consists of all standard bideterminants of the form* $(D \mid E)$ *where each row of* E *has the form* $1\ 2 \ldots p$ *for a suitable* p, $1 \leq p \leq n$.
Proof. These elements are linearly independent by Theorem 13.1 and U-invariant by Lemma 13.2. Thus, we need only show that they span $k[M_{m,n}]^U$. Suppose not, i.e., there are elements fixed by U which cannot be written as a linear combination of these elements. Let $j \geq 2$ be the smallest integer subject to the following condition: there is a U-invariant $F \in k[M_{m,n}]^U$ so that when F is written in terms of the standard basis, say $F = \Sigma c_i(D_i \mid E_i)$ with each $(D_i \mid E_i)$ a standard Young bitableau and each $c_i \neq 0$, then there is an s and a row in E_s where j is not preceded by $j - 1$.

Let $u = u_b \in U_{j-1,j}$ where $u_{j-1,j} = b$. Then as we noted in Lemma 13.2, $u{\cdot}x_{pq} = x_{pq}$ if $q \neq j$, $u{\cdot}x_{ij} = bx_{ij-1} + x_{ij}$, and u fixes every minor of the form $(\ldots \mid \ldots j-1\ j \ldots)$.

Now, let F be as above and let us write $F = \Sigma c_s(D_s \mid E_s) + \Sigma c_p'(D_p' \mid E_p')$ where (i) the c's are non-zero, (ii) there is at least one row in each E_s which contains j but not

$j - 1$, and (iii) if j appears in any row of E_p', then so does $j - 1$. As we have just seen, u fixes $\Sigma c_p'(D_p' \mid E_p')$.

For each E_s, the only possible occurrences of $j - 1$ and j in its rows are as follows: (1) j is preceded by $j - 1$, (2) $j - 1$ is followed by an integer larger than j, (3) $j - 1$ ends a row, (4) j is preceded by an integer smaller than $j - 1$ and (5) j starts a row. Furthermore, since E_s is standard, all rows of type (i) must occur above all rows of type (i+1). Let E_1 have the greatest number of rows, say m, of types (4) and (5). Applying u to F, we see that $c_1(D_1 \mid E_1)$ gives a term $b^m c_1(D_1 \mid E_1^*)$ where E_1^* is obtained from E_1 by replacing each j in rows of types (4) and (5) by $j - 1$; furthermore, E_1^* is standard. The tableau E_1^* uniquely determines E_1 for the following reasons. First, all rows of type (5) have been changed to rows beginning with $j - 1$; such rows must begin with j in E_1 by our choice of j. Otherwise, to obtain E_1, we change $j - 1$ to j in m rows of E_1^* reading from bottom to top. But, in $u \cdot F$, any other occurrence of $(D_1 \mid E_1^*)$ is with a coefficient cb^r, $r < m$. Since k is infinite, the coefficient of $(D_1 \mid E_1^*)$ depends on b. Therefore, F is not fixed by U. QED

Note. The group $GL(m,k)$ acts on $M_{m,n}$ by left translation. A proof just like that above shows that a basis of the algebra of U-invariants of this action consists of all standard bideterminants $(D \mid E)$ where each row of D has the form $i\ i{+}1 \ldots n$ for a suitable i.

Let $(D \mid E)$ be a bitableau and suppose that each row of E has the form $1\ 2 \ldots p$ for some p. Let E have e_i rows of length i for $1 \leq i \leq n$. Then, $(D \mid E)$ is U-invariant by Theorem 13.3 and if $t \in T$, then $t\,(D \mid E) = (e_1\omega_1 + \ldots + e_n\omega_n)(t)(D \mid E)$. Thus, the T-weight of $(D \mid E)$ is determined by the shape of $(D \mid E)$. We next derive some important consequences of Theorem 13.3.

Theorem 13.4 (the First Main Theorem). *Let* $G = SL(n,k)$. *Then* $R = k[M_{m,n}]^G$ *is generated by all minors of the form* $(i_1 \ldots i_n \mid 1\ 2 \ldots n)$.
Proof. We calculate the invariants of the Borel subgroup $B_1 = T_1U$ of G where T_1 is the subgroup of G consisting of all diagonal matrices of determinant 1. By Theorem 13.3 and the remarks preceding this proof, a basis for the space of B_1-invariants consists of all standard bitableaux $(D \mid E)$ where each row of E must be $1\ 2 \ldots n$ since $(D \mid E)$ is fixed by T_1. QED

Theorem 13.5. *For* $1 \leq i \leq n$, *let* e_i, e_i' *be non-negative integers. Let* $\omega = e_1\omega_1 + \ldots + e_n\omega_n$ *and* $\omega' = e_1'\omega_1 + \ldots + e_n'\omega_n$.
(a) *A basis for* $E(\iota\omega)$ *consists of all standard bitableaux* $(D \mid E)$ *where* E *has* e_i *rows of length* i *and each row of* E *has the form* $1\ 2 \ldots p$ *for a suitable* p.
(b) *The mapping from* $E(\omega) \otimes E(\omega')$ *to* $E(\omega + \omega')$ *given by* $\Sigma a \otimes b \to \Sigma ab$ *is surjective.*
Proof. We know that $k[GL_n] = k[M_{n,n}][1/\det]$ and that $E(\iota\omega) = \{f \in k[GL_n] : r_b f = \omega(b) f$ for all $b \in B\}$. Thus, statement (a) follows from Theorem 13.3 and the comments preceding Theorem 13.4. Statement (b) is an immediate consequence of (a). QED

We may now explain the notion of "GL_n-GL_m duality", a valuable concept in the representation-theoretic approach to invariant theory [50, 52]. Let V (resp. W) be a vector space over k having basis $v_1, \ldots, v_m$ (resp. $w_1, \ldots, w_n$). The group GL_m (resp.

GL_n) acts on V (resp. W) by $g \cdot v_j = \Sigma g_{ij} v_i$ (resp. $g \cdot w_j = \Sigma g_{ij} w_i$). Let $B_m = T_m U_m$ (resp. $B_n = T_n U_n$) be the standard Borel subgroup of GL_m (resp. GL_n), i.e., the subgroup consisting of all upper triangular matrices. The group $G = GL_m \times GL_n$ acts on $V \otimes W$ in the natural way: $(g_1, g_2)\cdot(v \otimes w) = g_1 v \otimes g_2 w$. We wish to give an explicit description of the algebra of $U_m \times U_n$ - invariants with respect to this action. This calculation is greatly facilitated by the following observation. The group G acts on the space $M_{m,n}$ by $(g, h)\cdot m = gm(h^T)$ and the mapping $V \otimes W \rightarrow M_{m,n}$ given by $\Sigma c_{ij} v_i \otimes w_j \rightarrow (c_{ij})$ is G-equivariant with respect to these actions. Thus, we may identify $k[V \otimes W]$ and $k[M_{m,n}]$ as G-modules.

Theorem 13.6 (GL_m-GL_n duality). *A basis for the $U_m \times U_n$ - invariant functions on $k[M_{m,n}]$ consists of all the standard bitableaux $(D \mid E)$ where each row has the form $(j\ j{+}1 \ldots m \mid n{-}m{+}j\ \ n{-}m{+}j{+}1 \ldots n)$.*

Proof. The actions of $GL_m \times \{e\}$ and $\{e\} \times GL_n$ commute so we may apply Theorem 13.3 and the note following it. We should also observe that if $u \in U_n$, then $u^T \in U_n^-$, the group of lower triangular matrices with 1's on the diagonal. Let s be the permutation $1 \rightarrow n$, $2 \rightarrow n - 1, \ldots, n \rightarrow 1$. Then $sU_n s^{-1} = U_n^-$ and if f is U_n-invariant, then sf is U_n^--invariant. QED

Note. In case char $k = 0$, knowledge of the $U_m \times U_n$ - invariants gives the decomposition of $k[V \otimes W]$ into irreducible G-modules. If char $k \geq 0$, then $k[M_{m,n}]$ is known to have a good filtration [32; Proposition 1.3d, p.722] and the $U_m \times U_n$ - invariants give its structure.

C. Results of K. Pommerening

The deepest results on finite generation questions for subgroups of the general linear group have been obtained by K. Pommerening. Our purpose in this section is to introduce his work. We should note at the beginning one change in convention. Whereas we have considered $k[G]^H = \{f \in k[G] : r_h f = f \text{ for all } h \in H\}$, Pommerening studies ${}^H k[G] = \{f \in k[G] : \ell_h f = f \text{ for all } h \in H\}$. The two algebras are isomorphic via the mapping ϵ: $k[G] \rightarrow k[G]$ given by $\epsilon(f)(g) = f(g^{-1})$, but we shall follow Pommerening and focus on ${}^H k[G]$.

As usual, let GL_n be the group of all invertible $n \times n$ matrices with coefficients in k. A *canonical unipotent subgroup* H of GL_n is given by a subset $\Psi \subset \{(i,j) : 1 \leq i < j \leq n\}$ with the property that if (i,j) and (j,p) are in Ψ, then $(i,p) \in \Psi$. The corresponding subgroup $H = U(\Psi)$ consists of all matrices $u = (u_{ij}) \in GL_n$ such that $u_{ij} = 1$ when $i = j$, u_{ij} is an arbitrary element in k when $(i,j) \in \Psi$, and $u_{ij} = 0$ otherwise. In the language introduced just before Lemma 3.11, the root system Ψ is closed and H is the unipotent radical of $G(\Psi)$. Alternatively, Ψ is "special" in the sense of [10; 14.5] and H is the corresponding unipotent group.

For H a canonical unipotent subgroup of GL_n, we wish to determine when ${}^H k[GL_n]$ is finitely generated over k. Now, $k[GL_n]$ is the localization of $k[M_{n,n}]$ at $\{\text{determinant}^i : i = 0,1,2,\ldots\}$ so ${}^H k[GL_n] = ({}^H k[M_{n,n}])_{\det}$. Thus, we are led to study ${}^H k[M_{n,m}]$ where GL_n acts on $M_{n,m}$ by left multiplication, $g \cdot m = gm$ and $(g \cdot f)(m) = f(g^{-1}m)$ for all $g \in GL_n$, $m \in M_{n,m}$, and $f \in k[M_{n,m}]$. (We shall always assume that $m \geq n$.) What facilitates this study is the presence of some easily described H-invariant functions, namely, certain minors.

Lemma 13.7. *Let $1 \leq i_1 < \ldots < i_p \leq n$ and $1 \leq j_1 < \ldots < j_p \leq m$. Then the minor $(i_1 \ldots i_p \mid j_1 \ldots j_p) \in {}^Hk[M_{n,m}]$ if and only if the following condition is satisfied: whenever $(i_r,q) \in \Psi$, then q appears among the $i_1, \ldots, i_p$.*

Proof. Let $(i,j) \in \Psi$ and let u_b be that element in H with 1's on the diagonal, b in entry (i,j), and 0's elsewhere. If $m \in M_{n,m}$, then $u_b{\cdot}m$ is that matrix obtained from m when its ith row is replaced by row i + $b{\cdot}$row j. Let $f = (i_1 \ldots i_p \mid j_1 \ldots j_p)$. If i does not appear among the i_r's, then $u_b{\cdot}f = f$. if i does appear, then $u_b{\cdot}(i_1 \ldots i \ldots i_p \mid j_1 \ldots j_p) = (i_1 \ldots i \ldots i_p \mid j_1 \ldots j_p) - b(i_1 \ldots j \ldots i_p \mid j_1 \ldots j_p)$. The second minor is 0 if and only if j appears among the i_r's. QED

A minor satisfying the condition in the previous lemma will be called a Ψ-minor. A bitableau $(S \mid T)$ is called a Ψ-bitableau if each of the minors corresponding to the rows of $(S \mid T)$ is a Ψ-minor. Pommerening conjectured that if H is a canonical unipotent subgroup of GL_n, then ${}^Hk[M_{n,m}]$ is generated by the (finitely many) Ψ-minors. This conjecture is still open. Pommerening did find evidence that it holds; for example, he proved that if ${}^Hk[M_{n,m}]$ is finitely generated over k, then it is integral over the algebra R generated by the invariant minors. He also showed that the localizations of ${}^Hk[M_{n,m}]$ and R at a certain minor are equal (Theorem 13.9(b)). We turn to that result next.

Lemma 13.8. *Let $y_1, \ldots, y_m$ be indeterminates over a k-algebra A, which is also an integral domain. Let the additive group $H = k^m$ act on the polynomial ring $A[y_1, \ldots, y_m]$ by: $h{\cdot}y_s = \Sigma_r h_r t_{rs} + y_s$ for $s = 1, 2, \ldots, m$ and $h = (h_1, \ldots, h_m) \in H$, where $T = (t_{rs})$ is an $m \times m$ invertible matrix with coefficients in A. Then $A[y_1, \ldots, y_m]^H = A$.*

Proof. For $f \in A[y_1, \ldots, y_m]$, let ∇f denote the gradient of f written as a column vector. Then, $(h_1, \ldots, h_m){\cdot}f - f = (h_1, \ldots, h_m)T\nabla f + \ldots + f((h_1, \ldots, h_m)T)$. All terms except the last contain the y-variables. Thus, if f is not in A we may find $(a_1, \ldots, a_m) = (h_1, \ldots, h_m)T$ such that the right-hand side is not zero. QED

Let $H = H(\Psi)$ be a canonical unipotent subgroup of GL_n. Let i be chosen, $1 \leq i \leq n - 1$. Suppose that $\{j : (i,j) \in \Psi\} = \{j_1, \ldots j_{m(i)}\}$. Let $d_i = (j_1 \ldots j_{m(i)} \mid j_1 \ldots j_{m(i)})$. Then d_i is a Ψ-minor. Indeed, by Lemma 13.7, we must show that if $(j_r,q) \in \Psi$, then q appears among the j's. But $(i,j_r) \in \Psi$ and $(j_r,q) \in \Psi$, so $(i,q) \in \Psi$ and q must be one of the j's by the definition of Ψ. Let $e_H = d_1 \ldots d_{n-1}$. Pommerening called e_H the "kritische Nenner" for H. In stating the next theorem, we take T as usual to be the subgroup of GL_n consisting of all diagonal matrices.

Theorem 13.9. *Let $H = H(\Psi)$ be a canonical unipotent subgroup of GL_n. Let R be the algebra generated over k by all the Ψ-invariant minors. Let $e_H \in R$ be the "kritische Nenner" for H.*

(a) $k[M_{n,m}][1/e_H] = R[1/e_H][x_{ij} : (i,j) \in \Psi]$;

(b) ${}^Hk[M_{n,m}][1/e_H] = R[1/e_H]$;

(c) *Let* $m = n$; *then,* $H = R' = \{g \in GL_n : \ell_g r = r \text{ for all } r \in R\}$;

(d) $H = \cap sUs^{-1}$ *where s runs over a certain subset of the Weyl group of T which includes* *the identity element.*

Proof. Let $r = \max\{i$: there is a j with $(i,j) \in \Psi\}$. We shall prove statements (a) and (b) by induction on r. Let $J = \{j : (r,j) \in \Psi\} = \{j_1, \dots, j_m\}$. Let $e_H = d_1 \dots d_r$ be the "kritische Nenner" for H; we put $d = d_r$.

Let K be the subgroup of H which has 0's in all of the (r,j_i) entries. Then, K is normal in H and $K = K(\Psi')$ where $\Psi' = \{(i,j) \in \Psi : i < r\}$. The "kritische Nenner" for K is $e_K = d_1 \dots d_{r-1}$. Let R (resp. S) be the subalgebra of $k[M_{n,m}]$ generated by all H-invariant (resp. K-invariant) minors. We now prove the theorem in several steps.

(i) $S = R[x_{r1}, \dots, x_{rm}]$. Obviously, $S \supset R[x_{r1}, \dots, x_{rm}]$. To prove the reverse containment, let $f \in S$, say, $f = (i_1 \dots i_p \mid -)$. If $f \in R$, we are finished. Otherwise, r must appear among the i's but some $j \in J$ must not. The minors obtained from f by deleting r from its rows and deleting any column are all in R. For suppose that $(i_q,r) \in \Psi$. Then $(i_q,r) \in \Psi$ and $(r,j) \in \Psi$ imply that $(i_q,j) \in \Psi$ which is not possible since $f \in S$. We now expand f by its rth row.

(ii) $S[1/d] = R[1/d][x_{rj} : j \in J]$. We localize both sides of (i) at d to see that it suffices to prove that $x_{rj} \in R[1/d][x_{rj} : j \in J]$. We choose an index j not in J. Then $(rj_1 \dots j_m \mid jj_1 \dots j_m) \in R$. We expand this minor by its rth row and obtain $x_{rj}d$ and other terms in $R[x_{rj} : j \in J]$.

(iii) Now we prove statement (a). We apply (ii) and the induction hypothesis to see that: $R[1/e_H][x_{ij} : (i,j) \in \Psi] = R[1/d][1/e_K][x_{ij} : (i,j) \in \Psi] = R[1/d][1/e_K][x_{ij} : (i,j) \in \Psi'][x_{rj} : j \in J] = S[1/d][1/e_K][x_{ij} : (i,j) \in \Psi'] = k[M_{n,m}][1/e_K][1/d] = k[M_{n,m}][1/e_H]$.

(iv) The elements x_{ij}, $(i,j) \in \Psi$ are algebraically independent over $Q(R)$, the quotient field of R. For let their transcendence degree over $Q(R)$ be p. Then, $p \leq \dim H$. But, by statement (a), Corollary 19.6(c) (whose proof does not require any of the material in this section) and the fact that $Q(R) \subset {}^H k(M_{n,m})$, we see that $nm = \dim Q(R) + p \leq nm - \dim H + p$. Thus, $p = \dim H$.

(v) The group H/K acts by left translation on $R[1/d][x_{rj} : j \in J]$ and

$$u \cdot x_{rj} = x_{rj} - \sum u_{rj_i} x_{j_i j}.$$

We now apply Lemma 13.8 (with $A = R[1/d]$) to see that the algebra of H/K-invariants is $R[1/d]$.

(vi) Now we prove statement (b). Using induction, (ii), and (v), we see that ${}^H k[M_{n,m}][1/e_H] = {}^{H/K}({}^K k[M_{n,m}][1/e_K][1/d]) = {}^{H/K}(S[1/e_K][1/d]) = {}^{H/K}(R[1/d][x_{rj} : j \in J][1/e_K]) = R[1/d][1/e_K] = R[1/e_H]$.

Next, we prove (c). We begin by noting that (c) holds for $H = U$ by Theorem 13.3 and the fact (Corollary 1.5) that U is observable in GL_n. Now, let $L = R'$. Then, $H \subset L \subset U$. Also, since R is sent to itself under left translation by T, we see that T normalizes L. Thus, L is a canonical unipotent subgroup [10; Proposition (2), p.184]. Let ${}^L k(GL_n)$ be the subfield of $k(GL_n)$ which consists of functions fixed under left translation by L. Then, $\dim {}^L k(GL_n) = \dim GL_n - \dim L$. However, ${}^H k(GL_n)$ is contained in $Q(R)$, the quotient field of R, by statement (b) and Theorem 2.1(5) so ${}^H k(GL_n) \subset Q(R) \subset {}^L k(GL_n)$, by assumption. But, obviously, ${}^L k(GL_n) \subset {}^H k(GL_n)$. We may conclude that $\dim H = \dim L$ and, then, that $H = L$.

Now, we prove statement (d). Let s be any permutation of $\{1,2,\dots,n\}$. We shall denote the corresponding permutation matrix by s, also; that is, s has 1's in the entries

$s{\cdot}i,i$ and 0's elsewhere. A direct calculation shows that $\ell_s{\cdot}(i_1 \dots i_p \mid j_1 \dots j_p) = (si_1 \dots si_p \mid j_1 \dots j_p)$. Suppose that H leaves invariant the minor $(i_1 \dots i_p \mid j_1 \dots j_p)$. Let s be the permutation of $\{1,2, \dots ,n\}$ defined by $s(n) = i_p, \dots ,s(n - p + 1) = i_1$ and, otherwise, s maps $\{1,2, \dots ,n - p\}$ to $\{1,\dots,\hat{\imath}_1,\dots,\hat{\imath}_p,\dots,n\}$ in an order preserving manner. Then, one may check using Lemma 13.7 that H fixes all minors of the form $\ell_s{\cdot}(i\ i{+}1 \dots n \mid \dots)$. Thus, for all $h \in H$, we have $\ell(s^{-1}hs){\cdot}(i\ i{+}1 \dots n \mid \dots) = (i\ i{+}1 \dots n \mid \dots)$ and $s^{-1}hs \in U$. On the other hand, let $L = \cap sUs^{-1}$, where s runs over all the permutations just defined. Then, L is observable, normalized by T, and fixes all of R. Thus, $L \subset R' = H$ by statement (c). QED

Pommerening used straightening to cancel the denominator e_H in Theorem 13.9(b) for certain $H(\Psi)$ and give conditions on Ψ equivalent to: the algebra ${}^Hk[M_{n,m}]$ is spanned by the standard Ψ-bitableaux [84; Theorem, p.288]. Namely, let $\Omega = \{1,2,\dots,n\}$. Two elements $\sigma,\tau \in \Omega$ are said to be *neighbors* if $(\sigma,\tau) \in \Psi$ and there is no element $\mu \in \Omega$ with (σ,μ) and $(\mu,\tau) \in \Psi$. Given $\sigma \in \Omega$, put $\Omega_\sigma = \{\tau \in \Omega : (\sigma,\tau) \in \Psi\}$. These sets give rise to an equivalence relation on Ω by defining σ to be related to τ if $\Omega_\sigma = \Omega_\tau$. Let the corresponding equivalence classes be denoted by M_i; for $\sigma \in M_i$, we put $\Omega_i = \Omega_\sigma$. The three required conditions are that the classes M can be numbered $M_1,\dots,M_r$ in such a way that (i) $\Omega_i \subset M_{i+1} \cup \dots \cup M_r$; (ii) $\Omega \supset \Omega_1 \supset \dots \supset \Omega_r = \varnothing$; (iii) for all i,j with $1 \le i < j < r$, we have: if more than one element of $M_{i+1} \cup \dots \cup M_r$ is not in Ω_i, then the elements of M_i have no neighbors in $M_{i+1},\dots,M_r$.

Applying these conditions, Pommerening showed that ${}^Hk[M_{n,m}]$ is finitely generated over k for a wide class of canonical unipotent subgroups including (i) unipotent radicals of parabolic subgroups and (ii) those subgroups obtained from type (i) by removing simple roots. However, in GL_5, consider the group H corresponding to $\Psi = \{(1,3), (1,4), (2,4), (2,5)\}$. Then (for example), $M_1 = \{1\}$, $M_2 = \{2\}$, $M_3 = \{3,4,5\}$, $\Omega_1 = \{3,4\}$, $\Omega_2 = \{4,5\}$, $\Omega_3 = \varnothing$ and condition (ii) can never be satisfied, i.e, the algebra ${}^Hk[M_{n,m}]$ is not spanned by the standard Ψ-bitableaux. However, we saw in §5 that the algebra of invariants is finitely generated. Consequently, one may show that the Popov-Pommerening conjecture holds for all GL_n where $n \le 5$ [109].

Exercises.

1. Let $r = \min\{m,n\}$. For $x \in M_{m,n}$ and $i = 1, \dots, r$, let $D_i(x)$ be the determinant of the $i \times i$ matrix obtained from x by choosing its first i rows and columns. The group $GL_m \times GL_n$ acts on the space $M_{m,n}$ by $(g,h){\cdot}x = gxh^{-1}$. Let U^- (resp. U^+) be the maximal unipotent subgroup of GL_m (resp. GL_n) consisting of all lower (resp. upper) triangular matrices with 1's on the diagonal and let $U = U^- \times U^+$. Show that the algebra of U-invariants on $M_{m,n}$ is generated by the D_i and is a polynomial algebra.

§14. The Relationship between A and $G{\cdot}A^U$

Let V be a rational G-module and let H be a subgroup of G. We denote by $G{\cdot}V^H$ the intersection of all the k-subspaces of V which contain V^H and which are invariant under G. In the case where G acts on a commutative k-algebra A, we let $G{\cdot}A^H$ be the intersection of all the k-subalgebras of A which contain A^H and which are invariant under

G. If A^H is finitely generated over k, say $A^H = k[f_1, \ldots, f_m]$, then (it is easily checked that) $G{\cdot}A^H$ is generated by the finite-dimensional subspaces $<G{\cdot}f_1>, \ldots, <G{\cdot}f_m>$.

Lemma 14.1. *Suppose that char $k = 0$. Let V be any rational G-module. Then $V = G{\cdot}V^U$. In particular, if V is a commutative k-algebra A, then A is finitely generated over k if and only if A^U is.*

Proof. Let $W = G{\cdot}V^U$. Since the actions of G are completely reducible, $V = W \oplus Z$, where Z is a G-invariant subspace of V. (Indeed, choose Z to be maximal among all G-invariant subspaces of V so that $W \cap Z = \{0\}$. Suppose that there is a $v \in V$ which is not in $W + Z$. Then $<G{\cdot}v>$ is a direct sum of irreducible G-modules, by complete reducibility. Let V' be one of these components. Either V' is contained in $W + Z$ or $V' \cap (W + Z) = \{0\}$. In the latter case, we replace Z by $Z + V'$.) If $Z \neq \{0\}$, then $Z^U \neq \{0\}$; but $Z^U \subset V^U \subset W$, contradicting the fact that $W \cap Z = \{0\}$. Thus, $Z = \{0\}$ and $V = W$.

Next, suppose that $V = A$ is a commutative k-algebra. If A is finitely generated over k, then so is A^U according to Theorem 9.4. On the other hand, suppose that $A^U = k[f_1,\ldots,f_m]$. The algebra $G{\cdot}A^U$ is finitely generated over k, being generated by the finite-dimensional subspaces $<G{\cdot}f_1>,\ldots,<G{\cdot}f_m>$. But, by the argument above, $A = G{\cdot}A^U$. QED

We now turn to the case char $k \geq 0$. The following result of Zariski and Samuel [118; Lemma, p. 198] plays a key role.

Theorem 14.2. *Let $k[x_1,\ldots,x_n]$ be the polynomial algebra over k in n indeterminates. Let M be the ideal generated by $x_1,\ldots,x_n$. Let $z_1,\ldots,z_m$ be homogeneous elements in M and let J be the ideal generated by $z_1,\ldots,z_m$. If the radical of J is M, then $k[x_1,\ldots,x_n]$ is integral over $k[z_1,\ldots,z_m]$.*

Proof. Let $S = k[x_1,\ldots,x_n]$ and let $R = k[z_1,\ldots,z_m]$. For each $i = 1,\ldots,n$, there is a positive integer $e(i)$ such that $x_i^{e(i)}$ is in J. Let $e = e(1) + \ldots + e(n)$; then M^e is contained in J. Indeed, let $f \in M$, say $f = s_1x_1 + \ldots + s_nx_n$ where $s_i \in S$. Then f^e is a sum of monomials $c(s_1x_1)^{a(1)} \ldots (s_nx_n)^{a(n)}$ where $c \in \mathbf{Z}$ and $a(1) + \ldots + a(n) = e$. Then, for at least one of the $a(i)$ we have $a(i) \geq e(i)$, i.e. $x_i^{a(i)} \in J$; hence, $f \in J$. Let $\{y_j\}$ be the finite set of monomials $x_1^{e(1)} \ldots x_n^{e(n)}$ where $e(1) + \ldots + e(n) < e$. We shall show that each monomial m in the x_i's can be written in the form $\Sigma_j r_j y_j$ where $r_j \in R$. It will follow that S is a finite module over R; then, the ring S is integral over R [71; Theorem 9.1, p.64].

We proceed by induction on the degree of m; if deg $m < e$, then m is one of the y_j's. So, let us suppose that deg $m \geq e$; we may then write $m = m'm''$ where m' and m'' are monomials with deg $m' = e$ and deg $m'' =$ deg $m - e$. The monomial m' is in M^e which is contained in the ideal J. Thus, m' may be written as $\Sigma_i s_i z_i$ where each s_i is an element of S, homogenous of degree deg m' - deg z_i. Now, $m = m'm'' = \Sigma_i (m''s_i)z_i$ where deg $m''s_i =$ deg m'' + deg m'- deg z_i = deg m - deg z_i. By the induction hypothesis, we may write $m''s_i$ as $\Sigma_j r_{ij}'y_j$ where each r_{ij}' is in R. Then, $m = m'm'' = \Sigma_i (m''s_i)z_i = \Sigma_{ij} (r_{ij}'z_i)y_j$ and the proof is finished. QED

Theorem 14.3. *The algebra A is integral over $G{\cdot}A^U$.*

Proof. We first prove the theorem in the case where $A = k[V]$, with V being a finite-dimensional G-module. Let $k[V]_d$ be the set of all polynomials in $k[V]$ which are homogeneous of degree d. Since G acts linearly on V, each of the subspaces $k[V]_d$ is invariant under G. It follows that if H is any subgroup of G and if $f \in k[V]^H$, then f can be written as a sum $\Sigma_d f_d$ where $f_d \in (k[V]_d)^H$. Let M be the ideal in $k[V]$ generated by all the $k[V]_d$, $d \geq 1$.

We shall denote the algebra $G{\cdot}k[V]^U$ by S. The algebra $k[V]^U$ is finitely generated over k by Theorem 9.4, say, $k[V]^U = k[f_1,\ldots,f_m]$ where f_i is homogeneous of degree $d_i > 0$. Then, the G-invariant algebra S is generated by (finitely many) homogeneous elements each of the form $g{\cdot}f_i$ for some $g \in G$. Let $S_d = k[V]_d \cap S$ and let J be the ideal in $k[V]$ generated by all the S_d, $d \geq 1$.

We now show that the radical of J is M. Indeed, let I be the radical of J in M. Since S is G-invariant, so is each S_d and the ideal J. Thus, I (and, of course, M) is G-invariant. Suppose that I is a proper subset of M. Then, there is an element $f + I \in (M/I)^U$ so that $f + I \neq I$. According to Theorem 9.9, there is an $F \in k[V]^U$ and a positive integer t so that $f^t - F \in I$. By definition, $F \in M^U \subset J$. Hence, $f^t \in I$ and, so, $f \in I$ which is a contradiction. It follows that $I = M$. We now apply Theorem 14.2 to conclude that $k[V]$ is integral over $G{\cdot}k[V]^U$.

Next, let A be finitely generated over k. There is a finite-dimensional vector space V on which G acts rationally and a surjective G-equivariant homomorphism π from $k[V]$ to A. Indeed, we may suppose that $A = k[a_1,\ldots,a_m]$ where the a_i's are linearly independent and $g{\cdot}a_j = \Sigma_i t_{ij}(g)a_i$ for suitable $t_{ij} \in k[G]$. Let $x_1,\ldots,x_m$ be indeterminates over k and let V^* have basis $\{x_1,\ldots,x_m\}$. We define $g{\cdot}x_j = \Sigma_i t_{ij}(g)x_i$, and $\pi(x_i) = a_i$. Since π is G-equivariant, $\pi(k[V]^U)$ is contained in A^U. Then, since π is surjective and $k[V]$ is integral over $G{\cdot}k[V]^U$, we may conclude that A is integral over $G{\cdot}A^U$.

Finally, we prove the general case. Let A be any commutative k-algebra on which G acts rationally. Let $f \in A$. Let B denote the smallest k-subalgebra of A which contains f and is invariant under G. Then B is the subalgebra generated by the finite-dimensional vector space $<G{\cdot}f>$. In particular, B is finitely generated over k and, by what was just proved, B is integral over $G{\cdot}B^U$, a subalgebra of $G{\cdot}A^U$. Hence, f is integral over $G{\cdot}A^U$. QED

We have seen, in Lemma 14.1, that $A = G{\cdot}A^U$ when char $k = 0$. However, as the example below shows, in general $G{\cdot}A^U$ is a proper subset of A.

Example. We follow the development and notation given at the beginning of §10 for $G = SL(2,k)$. Suppose that char $k = p > 0$ and let $V = V_p$ be the subspace of $k[G]^U = k[x_{11},x_{21}]$ spanned by all homogeneous polynomials of degree p. Let $A = S(V)$ be the symmetric algebra on V. We shall show that A strictly contains $G{\cdot}A^U$.

Let $e_i = x_{11}{}^i x_{21}{}^{p-i}$, $i = 0,\ldots,p$, be a basis for V. We have seen that ${}^UV = \{cx_{21}{}^p : c \in k\}$. Let $u \in U^-$, say $x_{21}(u) = b$. Then $\ell_u x_{21} = x_{21} - bx_{11}$. It follows that $\ell_u x_{21}{}^p = x_{21}{}^p - b^p x_{11}{}^p$, i.e., $\ell_u e_0 = e_0 - b^p e_p$. Thus, $<G{\cdot}e_0> = <U^-TU{\cdot}e_0> = <e_0,e_p>$ the space spanned by e_0 and e_p. In particular, $<G{\cdot}e_0> = G{\cdot}{}^UV$ is a proper subspace of V.

Let $S(V)_m$ be the subspace of $A = S(V)$ consisting of all polynomials homogeneous of degree m. The group G leaves invariant $S(V)_m$. Hence, $S(V)^U$ is a direct sum of the subspaces $(S(V)_m)^U$. This means that we may choose homogeneous elements $f_1,\ldots,f_r \in$

$S(V)^U$ such that $S(V)^U = k[f_1,\dots,f_r]$. Then $G{\cdot}S(V)^U$ is generated by the subspaces $<Gf_i>$. Since $<G{\cdot}e_o>$ is a proper subspace of V, we see that $G{\cdot}S(V)^U$ cannot contain $e_1,\dots,$ e_{p-1}.

Exercises.

1. Let A be a commutative k-algebra on which G acts rationally and let I be a G-invariant, prime ideal in A. Show that I is the radical of the ideal generated by $G{\cdot}I^U$.
2. Let A and B be commutative k-algebras on which G acts rationally and suppose that $B \subset A$. If A^U is integral over B^U, show that A is integral over B.
3. Let char $k = 0$. Prove the converse of Exercise 2.
4. Let char $k = 0$. Let P be a parabolic subgroup of G and let $H = \mathfrak{R}_u P$. Let A be a commutative k-algebra on which G acts rationally. Show that A is finitely generated over k if and only if A^H is.

§15. The Algebra grA

Let A be a commutative k-algebra on which the reductive G acts rationally. In this section, we construct a graded algebra gr$A$ which will play a central role in what follows. We maintain the notation of §3 throughout for G° and, in particular, fix a Borel subgroup $B = TU$ of G°. Furthermore, let $G = \cup_j g_j G^\circ$ be the coset decomposition of G with respect to G°. The group $g_jBg_j^{-1}$ is a Borel subgroup of G°. Thus, there is a $g \in G^\circ$ so that $g_jBg_j^{-1} = gBg^{-1}$. Then, the element $g^{-1}g_j$ is in the normalizer of B and $g^{-1}g_jG^\circ = g_jG^\circ$; we replace g_j by $g^{-1}g_j$ and assume from now on that each g_j normalizes both B and $U = B_u$. Furthermore, since the maximal tori in B are conjugate by elements in B, we may also assume that each g_j normalizes T. Then, the set of all elements in G of the form g_jt, where j is arbitrary and $t \in T$, forms a group. For $g_pg_q = g_rg$ for some r and $g \in G^\circ$; since the g_j's normalize U and T so does g which means that g is actually in T. Also, each g_j gives rise to a homomorphism of $X(T)$, $\chi \to g_j{\cdot}\chi$, defined by $(g_j{\cdot}\chi)(t) = \chi(g_j^{-1}tg_j)$. We may assume that the inner product on $\mathbf{R} \otimes_{\mathbf{Z}} X(T)$ is invariant under $N_G(T)$, the normalizer of T in G. Let $\alpha \in \Phi^+$; since $g_jUg_j^{-1} = U$, each $g_j{\cdot}\alpha$ is also a positive root and, in fact, $g_jU_\alpha g_j^{-1}$ is the corresponding one-dimensional subgroup.

We begin with a lemma essential to the construction of gr$A$. The key step will be to show that if γ is a simple root in Φ^+, then $\Sigma(\gamma,\alpha)/(\alpha,\alpha) = 1$ where the sum is over all positive roots α. A complete proof may be found in [12; Corollary, p.169]. We give an argument using basic facts about root systems listed in [54; appendix, pp.229-232].

Lemma 15.1. *There is a homomorphism h: $X(T) \to \mathbf{Z}$ such that*
(a) *$h(\omega)$ is a non-negative integer for every $\omega \in X^+(T)$;*
(b) *whenever $\chi, \chi' \in X(T)$ satisfy $\chi' > \chi$, then $h(\chi') > h(\chi)$;*
(c) *$h(g_j{\cdot}\chi) = h(\chi)$ for all j and all $\chi \in X(T)$.*

Proof. Suppose first that G is connected. We define $h(\chi) := \Sigma 2(\chi,\alpha)/(\alpha,\alpha)$ where the sum is over all positive roots α. Statement (a) follows from the definition of a dominant weight. To prove (b), we show that $h(\gamma) = 2$ for every simple root γ. For a moment, let Ψ be any abstract root system in a real vector space E. Let (α,β) denote an inner product on E relative to which the Weyl group of Ψ consists of orthogonal transformations. For $\alpha \in \Psi$, let τ_α denote the reflection relative to α which leaves Ψ invariant so $\tau_\alpha(\beta) = \beta - [2(\beta,\alpha)/(\alpha,\alpha)]\alpha$. Let Δ be a base for Ψ and let 2ρ be the sum of all the positive roots in Ψ. Let $\gamma \in \Delta$ and let α be a positive root. If $\alpha \neq \gamma$, then $\tau_\gamma(\alpha)$ is a positive root; if $\alpha = \gamma$, then $\tau_\gamma(\alpha) = -\alpha$. Hence, $\tau_\gamma(2\rho) = 2\rho - 2\gamma$. But, by definition, $\tau_\gamma(\rho) = \rho - [2(\rho,\gamma)/(\gamma,\gamma)]\gamma$. Thus, $2(\rho,\gamma)/(\gamma,\gamma) = 1$.

The vectors $\alpha^* = 2\alpha/(\alpha,\alpha)$, $\alpha \in \Psi$, form a root system in E called the dual of Ψ and denoted by Ψ^*; the positive roots in Ψ^* are the α^* where α is a positive root in Ψ (see the beginning of Section 6 in §1 of Chapter VI [12]). A basis for Ψ^* is Δ^*. We note that $2(\alpha^*,\beta^*)/(\beta^*,\beta^*) = 2(\beta,\alpha)/(\alpha,\alpha)$ for all $\alpha,\beta \in \Psi$. Now, applying the argument given in the preceding paragraph to Ψ^*, we see that if $2\rho^*$ is the sum of all the positive roots in Ψ^* and $\gamma \in \Delta$, then $1 = 2(\rho^*,\gamma^*)/(\gamma^*,\gamma^*) = \Sigma(\alpha^*,\gamma^*)/(\gamma^*,\gamma^*) = \Sigma(\gamma,\alpha)/(\alpha,\alpha)$ where the sum is over all positive roots α. This completes the argument when G is connected.

In general, we put $h(\chi) = \Sigma_j h^o(g_j \cdot \chi)$ where h^o is the homomorphism just constructed for the group G^o. If $\omega \in X(T)$ is dominant, then so is $g_j \cdot \omega$ since $g_j \cdot \alpha$ is a positive root whenever α is positive and $(g_j \cdot \omega, g_j \cdot \alpha) = (\omega, \alpha)$. The other properties follow immediately from the definition of h and what has been proved in the connected case. QED

Definition. For each integer n, let $A_n = \{a \in A : h(\chi) \leq n$ for all weights χ of T on $<G \cdot a>\}$.

Lemma 15.2. *Let A be a commutative k-algebra on which the reductive group G acts rationally.*

(a) *Each subspace A_n is G-invariant. If $n < 0$, then $A_n = \{0\}$. The G-invariant subspaces A_n give a G-invariant filtration of A, i.e., A is the union of all the A_n, $A_0 \subset A_1 \subset \ldots \subset A_n \subset \ldots$, and $A_nA_m \subset A_{n+m}$.*

(b) *If $v \in A_n$ and $t \cdot v = \chi(t)v$ for all $t \in T$ where $\chi \in X(T)$ and $h(\chi) = n$, then $v \in A_n^U$ and $v \notin A_{n-1}$.*

(c) *Let $v + A_{n-1}$ be a non-zero element in $(A_n/A_{n-1})^U$ such that $t \cdot v = \chi(t)v$ for some $\chi \in X(T)$. Then $v \in A_n^U$, $h(\chi) = n$, and χ is maximal among the weights of T on A_n/A_{n-1}.*

Proof. Let $a \in A_n$. Then $<G \cdot a> = <Gg \cdot a>$ for all $g \in G$ so $g \cdot a \in A_n$. Suppose next that $n < 0$ but $A_n \neq \{0\}$. Then there is a non-zero element v in A_n^U. Let $t \cdot v = \omega(t)v$; then $h(\omega) \geq 0$ by Lemma 15.1(a), a contradiction to the definition of A_n. Now, let us show that $A_nA_m \subset A_{n+m}$. Let $v \in A_n$ and $w \in A_m$. Then $<G \cdot vw> \subset <G \cdot v><G \cdot w>$. If χ (resp. χ') is a weight of T on $<G \cdot v>$ (resp. $<G \cdot w>$), then $h(\chi + \chi') = h(\chi) + h(\chi') \leq n + m$. The rest of statement (a) follows immediately from the definitions.

To prove statement (b), consider the subspace $<U \cdot v>$. Each weight ω of T on $<U \cdot v> \subset A_n$ has the form $\chi + \Sigma_i c_i \alpha_i$ where $c_i \in \mathbf{Z}$, $c_i \geq 0$, and $\alpha_i \in \Phi^+$ (by Lemma 3.1). Thus $\omega \geq \chi$ and $n = h(\omega) \geq h(\chi) = n$. It follows (from Lemma 15.1(b)) that $\omega = \chi$ and $<U \cdot v> = \{cv : c \in k\}$.

Next, we prove (c). Let $v + A_{n-1}$ be a non-zero vector in $(A_n/A_{n-1})^U$ such that $t \cdot v = \chi(t)v$ for some $\chi \in X(T)$. Suppose that every weight ω of T on $<U \cdot v>$ satisfies $h(\omega) \leq n - 1$. Then $<G^o \cdot v> = <U^- TU \cdot v> \subset A_{n-1}$ (where we have used Lemma 3.1 here). Applying Lemma 15.1(c), we see that $<G \cdot v> = \Sigma_j g_j \cdot <G^o \cdot v>$ is also contained in A_{n-1} and, so, $v \in A_{n-1}$ contradicting our assumption. Hence, there is a non-zero vector $v_o \in \langle U\cdot v\rangle \subset A_n$ such that $t\cdot v_0 = \chi(t)v_0$ and $h(\chi) = n$. It follows from (b) that v_0 is fixed by U. Now, v_0 is a linear combination of vectors $v + a$, $a \in A_{n-1}$. Since $h(\omega) = n$, we see by comparing T-weights that $v_0 = cv$, $c \in k^*$ and $v \in A_n^U$. Next, if $\omega' \in X(T)$ is a weight of T on A_n/A_{n-1} with $\omega' > \omega$, then $h(\omega') > h(\omega) = n$, contradicting the defining property of A_n. QED

We may now define a graded algebra gr$A$. As a vector space, gr$A$ is the direct sum of all the spaces A_n/A_{n-1}, $n = 0,1,\ldots$. If $a \in A_n$ and $b \in A_m$, then $ab \in A_{n+m}$ by Lemma 15.2(a); we define $(a + A_{n-1})(b + A_{m-1}) = ab + A_{n+m-1}$. In this way, gr$A$ becomes a commutative k-algebra on which G acts rationally. (Since gr$A$ depends on the homomorphism h defined in Lemma 15.1, we might write gr$_hA$, but this particular notation shall not be necessary in what follows.)

Lemma 15.3. *The algebras $(grA)^U$ and A^U are isomorphic and the isomorphism is T-equivariant.*

Proof. Let $v \in A^U$ with $t\cdot v = \chi(t)v$ and $h(\chi) = n$. We assign to v the element $v + A_{n-1}$ in $(\text{gr}A)^U$. This is an algebra isomorphism by Lemma 15.2, (b) and (c). QED

Next, let us consider the algebra $A^U \otimes k[G/U^-]$. The group T acts on this algebra by the natural action on A^U (since T normalizes U and, so, leaves A^U invariant) and by right translation on $k[G/U^-]$. The group G acts on this algebra by the trivial action on A^U and by left translation on $k[G/U^-]$. The actions of G and T commute. We put $R = (A^U \otimes k[G/U^-])^T$ along with the action of G just given.

For $\omega \in X^+(T)$, we define subspaces $Y(\omega)$ of $k[G/U^-]$ as follows: $Y(\omega) = \{f \in k[G/U^-] : r_tf = \omega(t)^{-1}f$ for all $t \in T\}$. If we choose a basis $\{a_i\}$ of A^U such that $t\cdot a_i = \omega_i(t)a_i$ for all $t \in T$, then we see that R is a direct sum of the subspaces $a_i \otimes Y(\omega_i)$. In case G is connected, $Y(\omega) = r(s_0)E(\omega)$ where $E(\omega)$ is as defined in §12 and s_0T is that element in the Weyl group of T such that $s_0Us_0^{-1} = U^-$. According to Theorem 12.1(b), $\dim {}^UY(\omega) = 1$. We let f_ω be a non-zero element in ${}^UY(\omega)$; then $\ell_tf_\omega = \omega(t)f_\omega$ and we may assume that $f_\omega(e) = 1$.

To study the relationship between the algebras R and gr$A$, we will construct a mapping Φ: gr$A \to R$. To this end, we define an ideal I in gr$A$ as follows. For each integer $n = 0,1,2,\ldots$, let I_n be the subspace of A_n/A_{n-1} spanned by all T-weight vectors having weight χ such that $h(\chi) < n$. Let I be the sum of all the I_n. The basic facts concerning I are as follows.

Lemma 15.4. (a) *The subspace I is an ideal in grA which is invariant under both U^- and T.*

(b) *As vector spaces, $grA = I \oplus (grA)^U$.*

(c) *There is a T-equivariant algebra homomorphism φ: $grA \to A^U$ such that $\varphi(u\cdot a) = \varphi(a)$ for all $u \in U^-$ and $a \in grA$.*

Proof. Let $a + A_{m-1} \in$ gr$A$ and $b + A_{n-1} \in I_n$. Suppose that $t\cdot a = \chi'(t)a$ and $t\cdot b = \chi(t)b$ for all $t \in T$. Then $h(\chi') \leq m$ and $h(\chi) < n$. The product $ab + A_{n+m-1}$ has T-weight $\chi' + \chi$ and $h(\chi' + \chi) = h(\chi') + h(\chi) < m + n$. This shows that I is an ideal in gr$A$. By definition, it is invariant under T. If $a + A_{n-1} \in I_n$ with $t\cdot a = \chi(t)a$ and if $u \in U^-$, then $u{\cdot}a$ is a sum of T-weight vectors having weights of the form $\chi - \Sigma_i c_i\alpha_i$ where $c_i \in \mathbf{Z}$, $c_i \geq 0$, and $\alpha_i \in \Phi^+$. But $h(\chi - \Sigma_i c_i\alpha_i) \leq h(\chi) < n$ by Lemma 15.1(b). This proves statement (a).

By definition, $A_n/A_{n-1} = I_n \oplus V_n$ where V_n is the subspace of A_n/A_{n-1} spanned by all T-weight vectors having weight χ such that $h(\chi) = n$. Thus, statement (b) follows from Lemma 15.2(b) and (c). We have already observed that the algebras $(\mathrm{gr}A)^U$ and A^U are isomorphic and that this isomorphism is T-equivariant. This leads to a mapping φ: $\mathrm{gr}A \to (\mathrm{gr}A)^U \simeq A^U$ by first projecting A to $(\mathrm{gr}A)^U$. Let $v \in \mathrm{gr}A$, say $v = a + b$ where $a \in I$ and $b \in (\mathrm{gr}A)^U$. Let $u \in U^-$. Then $u{\cdot}v = (u{\cdot}a + u{\cdot}b - b) + b$. The vector $u{\cdot}a + u{\cdot}b - b \in I$. Also, $\varphi(v) = \varphi(b) = \varphi(u{\cdot}v)$. QED

Let $\{b_i\}$ be any basis of the vector space A^U. A *morphism from G/U^- to A^U* is any mapping α: $G/U^- \to A^U$ of the form $\alpha(gU^-) = \Sigma_i f_i(g)b_i$ where each $f_i \in k[G/U^-]$ and where only finitely many of the functions f_i are non-zero. The algebra consisting of all morphisms from G/U^- to A^U admits three group actions. Let α be as above. Then, the group G acts by left translation, namely, $(g{\cdot}\alpha)(g_1U^-) = \alpha(g^{-1}g_1U^-) = \Sigma_i \ell_g f_i(g_1)b_i$. The group T acts by right translation, namely, $(t{\cdot}\alpha)(g_1U^-) = \alpha(g_1tU^-) = \Sigma_i r_t f_i(g_1)b_i$. The group T also acts by $(t * \alpha)(g_1U^-) = t{\cdot}\alpha(g_1U^-) = \Sigma_i f_i(g_1)t{\cdot}b_i$ since A^U is invariant under T.

The algebra $A^U \otimes k[G/U^-]$ also has three group actions: an action of G (resp. T) coming from the action of G (resp. T) on $k[G/U^-]$ by left translation (resp. right translation) and an action of T coming from the action of T on A^U. Furthermore, the vector space $A^U \otimes k[G/U^-]$ may be identified with the vector space consisting of all morphisms from G/U^- to A^U by associating to an element $\Sigma_i b_i \otimes f_i$ the morphism sending gU^- to $\Sigma_i f_i(gU^-)b_i$. It is easy to check that this identification is equivariant with respect to each of the three actions defined above.

Theorem 15.5. *Let Φ: $grA \to R = (A^U \otimes k[G/U^-])^T$ be defined by $\Phi(b)(gU^-) = \varphi(g^{-1}{\cdot}b)$ for all $g \in G$ and $b \in grA$. Then, Φ is an injective, G-equivariant algebra homomorphism. The algebra R is integral over (the image of) grA. If G is connected, then Φ gives a surjective isomorphism of $(grA)^U$ to R^U.*

Proof. If $b \in \mathrm{gr}A$, then $\Phi(b)$ defines a mapping from G/U^- to A^U. Indeed, for all $g \in G$ and $u \in U^-$, we have $\Phi(b)(gu) = \varphi(u^{-1}g^{-1}{\cdot}b) = \varphi(g^{-1}b)$ by Lemma 15.4(c). Since the action of G on $\mathrm{gr}A$ is rational, there are elements $b_1,\ldots,b_n \in \mathrm{gr}A$ and functions $t_j \in$ k[G] so that $g{\cdot}b = \Sigma_j t_j(g)b_j$ for all $g \in G$. Let $b_j = c_j + d_j$ where $c_j \in I$ and $d_j \in (\mathrm{gr}A)^U$. Then $\Phi(b)(gU^-) = \varphi(g^{-1}{\cdot}b) = \varphi(\Sigma_j t_j(g^{-1})(c_j + d_j)) = \Sigma_j t_j(g^{-1})\varphi(d_j)$. Thus we see that $\Phi(b)$ is a morphism from G/U^- to A^U. That $\Phi(b)$ is fixed under the action of T (i.e., corresponds to an element in R) follows from the T-equivariance of φ. Furthermore, an easy calculation shows that Φ is an algebra homomorphism, e.g., $\Phi(b_1b_2)(gU^-) = \varphi(g^{-1}{\cdot}b_1b_2) = \varphi(g^{-1}{\cdot}b_1)\varphi(g^{-1}{\cdot}b_2) = \Phi(b_1)(gU^-){\cdot}\Phi(b_2)(gU^-)$. Next, let us show that Φ is G-equivariant. If $g, g_1 \in G$ and $b \in \mathrm{gr}A$, then $(g{\cdot}\Phi(b))(g_1U^-) = \Phi(b)(g^{-1}g_1U^-) = \varphi(g_1{}^{-1}g{\cdot}b) = \Phi(g{\cdot}b)(g_1U^-)$.

Since Φ is G-equivariant, its kernel = ker Φ, is G-invariant. Suppose that ker $\Phi \neq \{0\}$. Then, there is a non-zero element $b \in (\mathrm{gr}A)^U \cap \ker \Phi$. But $\Phi(b)(eU^-) = \varphi(b) \neq 0$. Thus, Φ is injective.

Now, let us assume that G is connected and prove the last statement in the

Theorem. As above, let f_ω be the unique non-zero element in ${}^UY(\omega)$ with $\ell_t f_\omega = \omega(t)f_\omega$ and $f_\omega(e) = 1$. Then, the algebra R^U is spanned by elements $a \otimes f_\omega$ where $a \in A^U$ and $t \cdot a = \omega(t)a$. We note that $a \otimes f_\omega$ corresponds to the morphism α: $G/U^- \to A^U$ such that for $u_1 \in U^-$, $t \in T$, $u \in U$, we have $\alpha(utu_1U^-) = f_\omega(utu_1)a = \omega(t^{-1})a$. Now, $\Phi(a)(utu_1U^-) = \varphi(t^{-1}u^{-1}\cdot a) = \varphi(t^{-1}\cdot a) = \omega(t^{-1})a$ and, so, $\Phi(a) = a \otimes f_\omega$. This shows that Φ gives a surjective isomorphism of $(\text{gr}A)^U$ to R^U.

Finally, we show that R is integral over $\Phi(\text{gr}A)$ for an arbitrary (not necessarily connected) reductive group G. Let H be the group consisting of all elements of the form g_jt where j is arbitrary and $t \in T$. The group H normalizes U, U^- and T and, so, acts on A^U and also by right translation on $k[G/U^-]$. This means that the finite group H/T acts on R; this action commutes with the action of G on R (given by left translation on $k[G/U^-]$) and, so, sends R^U to R^U. We shall show that the H/T fixed points of this action are precisely $\Phi((\text{gr}A)^U)$. This will show that R^U is integral over $\Phi((\text{gr}A)^U)$. (In general, if a finite group G acts on a commutative k-algebra A, then A is integral over A^G. For an element $a \in A$ satisfies the equation $\Pi_{g\in G}(x - g\cdot a) = 0$ which has coefficients in A^G.) Then, according to Theorem 14.3, R is integral over $G{\cdot}R^U$ which in turn is integral over $G{\cdot}\Phi((\text{gr}A)^U). = \Phi(G{\cdot}(\text{gr}A)^U) \subset \Phi(\text{gr}A)$.

So, let us show that $\Phi((\text{gr}A)^U)$ is the algebra of invariants for H/T acting on R^U. Let $a \in A^U = (\text{gr}A)^U$ with T-weight ω, say. Let $u_1 \in U^-$, $t \in T$, and $u \in U$. Then $\Phi(a)(g_iutu_1) = \varphi(u_1^{-1}t^{-1}u^{-1}g_i^{-1}\cdot a) = (g_i^{-1}\cdot\omega)(t^{-1})g_i^{-1}\cdot a$ where we have used Lemma 15.4 and the fact that $g_i^{-1}\cdot a$ is fixed by U. Now, we study the action of H/T on R^U. To begin, we note that $R = (A^U \otimes k[G/U^-])^T = \oplus_j(A^U \otimes \ell(g_j)k[G^\circ/U^-])^T$ and, so, $R^U = \oplus_j(A^U \otimes \ell(g_j)^Uk[G^\circ/U^-])^T$. Let $a \in A^U$ and $f_\omega \in k[G^\circ/U^-]$ be as above. By definition, $g_i^{-1}\cdot(a \otimes f_\omega) = g_i^{-1}\cdot a \otimes r(g_i^{-1})f_\omega$. A calculation shows that $r(g_i^{-1})f_\omega(g_jutu_1)$ is 0 unless $j = i$ in which case we get $(g_i^{-1}\cdot\omega)(t^{-1})$. Therefore, $\Sigma_i\, g_i^{-1}\cdot(a \otimes f_\omega) = \Phi(a)$. Since any H/T invariant is a linear combination of elements of the form $\Sigma_i\, g_i^{-1}\cdot(a \otimes f_\omega)$, we are finished. QED

Note. Suppose that G is connected. The mapping Φ: $\text{gr}A \to R$ has been shown to be an injective, G-equivariant algebra homomorphism. When is Φ surjective? First, if char $k = 0$, then Φ is surjective since we have shown that $(\text{gr}A)^U = R^U$. If char $k > 0$, the situation is more complicated. In fact, it may be shown that Φ is surjective if and only if the algebra A has a "good filtration" [42; Theorem 16, p. 133]. By this we mean the following. Let Z be a locally finite G-module (having a countable basis). Then Z has a *good filtration* if there is a sequence of G-invariant subspaces $0 = Z_0 \subset Z_1 \subset \ldots$ such that $Z = \cup Z_i$ and each Z_i/Z_{i-1} is isomorphic to some $E(\omega_i)$. If char $k = 0$, then all such Z have good filtrations since actions of G are completely reducible. If char $k > 0$, then locally finite G-modules may or may not have good filtrations. In particular, let $G \times G$ act on G by $(a,b)\cdot x = axb^{-1}$ and consequently on $k[G]$. It is known that $k[G]$ has a good filtration with respect to this action [30, p.72]. Let us construct R and $\text{gr}k[G]$ as above. We observed following Lemma 15.3 that, quite generally, R is a direct sum of subspaces $a_i \otimes Y(\omega_i)$. In the case at hand, a_i is an element in $k[G]$ fixed by $U \times U$ and with $T \times T$ weight (ω', ω) say. By Theorem 12.1(b), $\omega' = \iota\omega$. It follows that $\text{gr}k[G] = R = \Sigma_\omega E(\iota\omega) \otimes E(\omega)$.

We next look at some geometric properties of $\text{gr}A$ and R, especially as they relate to the structure of A. If I is a G-invariant ideal in A, we define $I(R)$ to be $(I^U \otimes k[G/U^-])^T$. Then, $I(R)$ is a G-invariant ideal in R.

Lemma 15.6. (a) *Let I be a G-invariant radical ideal in A. Then, $I(R)$ is a G-invariant radical ideal in R. If G is connected and if I is a prime ideal in A, then $I(R)$ is a prime ideal in R.*
(b) *Let I and J be G-invariant radical ideals in A and suppose that I is a proper subset of J. Then, $I(R)$ and $J(R)$ are G-invariant radical ideals in R and $I(R)$ is a proper subset of $J(R)$.*
Proof. The exact sequence $0 \to I^U \to A^U \to A^U/I^U \to 0$ gives rise to another exact sequence $0 \to I^U \otimes k[G/U^-] \to A^U \otimes k[G/U^-] \to A^U/I^U \otimes k[G/U^-] \to 0$. This exact sequence in turn leads to the following sequence which is also exact (see Exercise 6, §11): $0 \to (I^U \otimes k[G/U^-])^T \to (A^U \otimes k[G/U^-])^T \to (A^U/I^U \otimes k[G/U^-])^T \to 0$. If I is a G-invariant radical ideal in A, then neither of the algebras A^U/I^U or $k[G/U^-]$ has nilpotent elements, so their tensor product also does not have nilpotent elements (Exercise 1). Thus, the subalgebra $(A^U/I^U \otimes k[G/U^-])^T$ does not have nilpotent elements which means that $I(R)$ is a radical ideal in R. If G is connected and I is a prime ideal in A, then neither A^U/I^U nor $k[G/U^-]$ has zero divisors so their tensor product also does not have zero divisors (Exercise 1).

To prove (b), it is enough to show that $I(R)$ is a proper subset of $J(R)$. Now, the U-invariant elements in $I(R)$ are sums of elements having the form $a \otimes {}^UY(\omega)$ where $a \in I^U$ and $t{\cdot}a = \omega(t)a$ for all $t \in T$. Then, $I(R)^U$ is a proper subset of $J(R)^U$ since I^U is a proper subset of J^U (Exercise 4, §9). Then, $I(R)$ is a proper subset of $J(R)$. QED

Let X be an affine variety on which G acts morphically and let $A = k[X]$. The algebra A^U is finitely generated over k by Theorem 9.4 so $R = (A^U \otimes k[G/U^-])^T$ is a finitely generated k-algebra. Let Y be the corresponding affine variety.

Corollary 15.7. (a) *To each closed G-invariant subvariety Z in X there corresponds a closed G-invariant subvariety in Y; this correspondence is order preserving.*
(b) *The stabilizer of each point in Y contains a maximal unipotent subgroup.*
Proof. Statement (a) follows from Lemma 15.6 since G-invariant subvarieties correspond to G-invariant radical ideals. Now, let Z be the affine variety such that $k[Z] = k[G/U^-]$ and let $z \in Z$ be a point such that $G_z = U^-$ (Theorem 4.3). Suppose that Z is contained in a finite-dimensional rational G-module V and that $z = v_1 + \ldots + v_r$ is the decomposition of z into a sum of non-zero T-weight vectors in V having distinct T-weights. Since U^- fixes z, it fixes each v_i (Lemma 3.8). According to Theorem 5.3, $Z = \mathrm{cl}(G{\cdot}z) = \cup_j g_j \mathrm{cl}(G^\circ{\cdot}z) = \cup g_j G^\circ{\cdot}\mathrm{cl}(B^-{\cdot}z) = G{\cdot}\mathrm{cl}(B^-{\cdot}z)$. However, $B^-{\cdot}z = TU^-{\cdot}z = T{\cdot}z$. Hence, each element in $\mathrm{cl}(B^-{\cdot}z)$ has the form $c_1v_1 + \ldots + c_rv_r$ for certain $c_i \in k$ and U^- is contained in its stabilizer. Since the actions of G and T on R commute, the stabilizer of each point in Y also contains a maximal unipotent subgroup. QED

There is a close geometric relationship between the actions of G on A and on gr$A$: namely, the action of G on Spec A *contracts* to the action of G on Spec gr$A$. We make this idea precise in the rest of this section.

Let x be an indeterminate and let $A[x]$ be the polynomial algebra over A. Let D

be the sum of all the $A_n x^n$, $n = 0,1,\ldots$. We denote the localization of $A[x]$ (resp. D) at the set $\{1,x,x^2,\ldots\}$ by $A[x,1/x]$ (resp. $D[1/x]$). We note that $D[1/x] = A[x,1/x]$, a fact which will be used several times below.

The action of G on A gives rise to a rational action of G on $A[x]$ be defining $g{\cdot}x = x$ for all $g \in G$. The group k^* also acts rationally on $A[x]$ by defining $c{\cdot}x = cx$ for all $c \in k^*$.

Lemma 15.8. *The algebra D/xD is isomorphic to grA and this isomorphism is G-equivariant.*

Proof. Let π_i: $A_i \to A_i/A_{i-1}$ be the canonical maps for $i \geq 0$ and define a mapping ψ: $D \to grA$ by $\psi(\Sigma_i a_i x^i) = \Sigma_i \pi_i(a_i)$. The mapping ψ is (obviously) surjective, k-linear, and G-equivariant. It is also an algebra homomorphism since for $a \in A_n$, $b \in A_m$, we have $\psi(ax^n)\psi(bx^m) = (a + A_{n-1})(b + A_{m-1}) = ab + A_{n+m-1} = \psi(ax^n{\cdot}bx^m)$. Now let $f = \Sigma_i a_i x^i$ be in D. Then, $\psi(f) = 0$ if and only if each a_i is in A_{i-1}. Thus, $\psi(f) = 0$ if and only if $f \in xD$. QED

We shall be using a number of facts from commutative algebra. In stating these, we will follow [71]. Let B be any commutative ring with identity. The set of all prime ideals in B is called the *spectrum* of B and is denoted by Spec B. If I is any ideal in B, the set $V(I)$ is defined to be $\{P \in \text{Spec } B : P \supset I\}$. There is a topology on Spec B, called the *Zariski topology*, for which the closed sets are precisely the $V(I)$'s. If $P \in$ Spec B, the quotient field of B/P will be denoted by $\kappa(P)$.

Let f: $B \to C$ be a ring homomorphism which takes the identity in B to the identity in C. The mapping from Spec C to Spec B given by $Q \to f^{-1}(Q)$ is continuous in the Zariski topology. If $P \in$ Spec B, the *fiber of f over P* is defined to be Spec $(C \otimes_B \kappa(P))$. In addition to these definitions, the following theorems will be used.

Lemma 15.9 [71; Exercise 11.8, p.86]. *A module over a Dedekind ring is flat if and only if it is torsion free.*

Lemma 15.10 [71; Example 1, p.269]. *Let B and C be any commutative rings with identities and let f: $B \to C$ be a ring homomorphism. If I is an ideal in B, then $C \otimes_B (B/I) = C/IC$.*

Lemma 15.11 [71; Theorem 4.2, p.23]. *Let B be any commutative ring with identity. Let S be a multiplicative set in B, let I be an ideal in B, and let S' be the image of S in B/I. Then, the ring B_S/IB_S is isomorphic to B/I localized at S'.*

Lemma 15.12 [71; Corollary 4, p.24]. *Let B be any commutative ring with identity and let $P \in$ Spec B. If S is any multiplicative set in B disjoint from P, then*

$$B_P = (B_S)_{PB_S}.$$

Lemma 15.13. *Let $\alpha \in k$ and let M be a maximal ideal in $A[x]$ containing $x - \alpha$. Let $V(I)$ be a closed set in Spec $A[x]$ where I is an ideal in $A[x]$. Suppose that $V(I)$ is k^*-invariant and contains M. Then there is a maximal ideal in $V(I)$ which contains x.*

Proof. We may assume that $\alpha \neq 0$. Each element in $A[x]$ may be written uniquely in the form $a_0 + a_1(x-\alpha) + \ldots + a_n(x-\alpha)^n$ where each a_i is in A. Such an element is in M if and only if a_0 is. In particular, $A[x]/M = A/(M \cap A)$ and $M \cap A$ is maximal in A. The ideal $c{\cdot}M$, $c \in k^*$, is the set of all elements in $A[x]$ having the form $a_0 + a_1(x - \alpha') + \ldots + a_n(x - \alpha')^n$ where each $a_i \in A$, $a_0 \in M$, and $\alpha' = \alpha/c$. Let $i(x) \in I$ where $i(x) = i_0 + i_1x + \ldots + i_nx^n$, $i_j \in A$. The canonical homomorphism $A \to A/(M \cap A)$ extends to a homomorphism from $A[x]$ to $(A/(M \cap A))[x]$. Let $i'(x)$ be the image of $i(x)$ under this mapping. Since $i(x) \in c{\cdot}M$, we have $i(x) = a_0 + a_1(x - \alpha') + \ldots + a_n(x - \alpha')^n$ where $a_0 \in M \cap A$ and $\alpha' = \alpha/c$. Therefore, $i'(\alpha') = 0$ and, since k is infinite, we see that $i'(x) = 0$. Hence, $i(x) \in (M \cap A)[x]$. It follows that I is contained in the maximal ideal generated by $M \cap A$ and x. QED

Theorem 15.14. *Let i be the inclusion of* $k[x]$ *in D. Then D is flat over* $k[x]$. *If* $M = (x) \subset k[x]$ *(resp.* $(x - \alpha)$, $\alpha \in k^*$*), then the fiber of i over M is Spec (grA) (resp. Spec A).*

Proof. That D is flat over $k[x]$ is immediate using Lemma 15.9. Next, let $M = (x)$. By definition, the fiber of i over M is Spec $(D \otimes_{k[x]} k[x]/M)$. But this is Spec D/xD by Lemma 15.10 which is Spec (gr$A$) according to Lemma 15.8.

Now, let $\alpha \in k^*$ and let $M = (x - \alpha) \subset k[x]$. We need to show that Spec$(D \otimes_{k[x]} k[x]/M) =$ Spec A. By Lemma 15.10, $D \otimes_{k[x]} k[x]/M = D/MD$. Since x is a unit in D/MD, we have $D[1/x]/MD[1/x] = D/MD$ by Lemma 15.11. Then, $D/MD = A[x,1/x]/MA[x,1/x] = A[x]/MA[x]$ again using Lemma 15.11. Finally, it is easily seen that each element in $A[x]$ may be written uniquely in the form $a_n(x - \alpha)^n + \ldots + a_1(x - \alpha) + a_0$ where each $a_i \in A$. Thus, $A[x]/MA[x] = A$. QED

In the context of Theorem 15.14, we say that the action of G on A *contracts* to the action of G on gr$A$ and that the action of G on A is a *deformation* of the action of G on gr$A$.

Exercises.

1. Let A and B be k-algebras on which G acts rationally. If neither A nor B has nilpotent elements (resp. zero divisors), then $A \otimes B$ has no nilpotents (resp. zero divisors). (Hint: reduce to the case where A and B are finitely generated by replacing A by the algebra generated by $<G{\cdot}a>$.)
2. Let G be connected. Let A be a finitely generated k-algebra having no zero divisors and let I be a G-invariant ideal in A. Let R and $I(R)$ be as constructed in this section. Show that dim $R/I(R)$ = dim A/I. (Hint: show that up to integral closure, $R/I(R)$ is gr A/I.)

§16. Finite Generation and *U*-invariants

We now apply the theory in §15 to the problem of finite generation. In this section, we retain the notation and terminology of §12.

A. Algebras

Lemma 16.1. *If grA is finitely generated over k, then so is A.*
Proof. Suppose that gr$A$ is generated by elements $f_i + A_{n(i)-1}$ for $i = 1,\ldots,r$. We shall show that $A = k[f_1,\ldots,f_r]$; in fact, we shall show by induction on n that $A_n \subset k[f_1,\ldots,f_r]$. The case $n = 0$ is immediate since $A_{-1} = \{0\}$. Now, suppose that $n > 0$ and that for all $m < n$, we have $A_m \subset k[f_1,\ldots,f_r]$. Let $f \in A_n$. There is a polynomial p in $f_1,\ldots,f_r$ so that $f - p(f_1,\ldots,f_r) \in A_{n-1}$. The induction hypothesis now shows that $f \in k[f_1,\ldots,f_r]$. QED

Theorem 16.2. *Let A be a commutative k-algebra on which the reductive group G acts rationally. Then A is finitely generated over k if and only if A^U is.*
Proof. We may assume that G is connected. If A is finitely generated over k, then so is A^U by Theorem 9.4. So, let us assume that A^U is finitely generated over k. In §15 (Lemma 15.3 and Theorem 15.5), we showed that the algebras A^U, $(\mathrm{gr}A)^U$, and R^U are all isomorphic; thus, each is finitely generated over k. It then follows that the algebras $G{\cdot}(\mathrm{gr}A)^U$ and $G{\cdot}R^U$ are finitely generated over k. Furthermore, the algebra $R = (A^U \otimes k[G/U^-])^T$ is finitely generated over k (by Theorem A) since both A^U and $k[G/U^-]$ are. The mapping Φ: gr$A \rightarrow R$ is an injective, G-equivariant algebra homomorphism by Theorem 15.5; we shall identify gr$A$ with its image in R. According to Theorem 14.3, R is integral over $G{\cdot}R^U = G{\cdot}(\mathrm{gr}A)^U$. Since R itself is finitely generated over k, R is a finitely generated $G{\cdot}(\mathrm{gr}A)^U$ module. (Indeed, suppose that $R = k[f_1,\ldots,f_r]$ and that $f_i^{e(i)} + b_1 f_i^{e(i)-1} + \ldots + b_{e(i)} = 0$ for $b_j \in G{\cdot}(\mathrm{gr}A)^U$. Then, R is generated over $G{\cdot}(\mathrm{gr}A)^U$ by all monomials $f_1^{a(1)} \ldots f_r^{a(r)}$ where $a(i) < e(i)$.) Hence, the submodule gr$A$ is also a finitely generated $G{\cdot}(\mathrm{gr}A)^U$ module. Since $G{\cdot}(\mathrm{gr}A)^U$ is finitely generated over k, so is gr$A$. Then A is finitely generated over k by Lemma 16.1. QED

Theorem 16.3. *Let H be a subgroup of G. Let L be a reductive subgroup of G which is contained in the normalizer of H in G. Let U(L) be a maximal unipotent subgroup of L. Let G act rationally on a commutative k-algebra A. Then A^H is finitely generated over k if and only if $A^{HU(L)}$ is.*
Proof. Since L normalizes H, the algebra A^H is invariant under L. We apply Theorem 16.2. QED

J. Horvath has used Theorem 16.3 in the case where H is obtained from the unipotent radical of a parabolic subgroup by removing simple roots [49].

Theorem 16.4. *Let H be the unipotent radical of a parabolic subgroup of G^o.*
(a) *Then H is a Grosshans subgroup of G.*
(b) *If A is finitely generated over k, then so is A^H.*
Proof. We may assume that G is connected (Corollary 4.4). Let H be the unipotent radical of the parabolic subgroup P and let $P = LH$ be the Levi decomposition of P, where L is reductive. Let $U(L)$ be a maximal unipotent subgroup of L. Then $U = U(L)H$ is a maximal unipotent subgroup of G. The algebra A^U is finitely generated over k by Theorem 9.4. Since $A^U = (A^H)^{U(L)}$, we may apply Theorem 16.3 and Corollary 1.5. QED

Theorem 16.4 was first proved in the case char $k = 0$ in [47]. A characteristic free proof using representation theory is given in [31]. Another characteristic free proof, using an explicit codimension 2 embedding, is given in [39].

In Section 15, we defined the algebra D to be the sum of all the $A_n x^n$ for $n = 0,1,\ldots$.

Lemma 16.5. *If A is finitely generated over k, then so are gr$A$, R, and D.*

Proof. According to Lemma 15.3, $A^U = (\mathrm{gr}A)^U$. Thus, if A is finitely generated over k, then so are A^U and gr$A$ by Theorem 16.2. Also, $R = (A^U \otimes k[G/U^-])^T$ is finitely generated over k. To show that D is finitely generated over k, we need only show that D^U is. Let A^U have generators $a_1,\ldots,a_r$ where we may assume that $t \cdot a_i = \omega_i(t)a_i$ for all $t \in T$ and $h(\omega_i) = n(i)$. We shall show that D^U is generated by x and all the $a_i x^{n(i)}$. Let $ax^n \in D^U$ where $a \in A_n^U$, $t \cdot a = \omega(t)a$ and $h(\omega) \leq n$. Then, a is a linear combination of monomials $a_1^{d(1)} \ldots a_r^{d(r)}$ where the $d(j)$ are non-negative integers and $\omega = d(1)\omega_1 + \ldots + d(r)\omega_r$. Hence, ax^n is a linear combination of monomials $(a_1 x^{n(1)})^{d(1)} \ldots (a_r x^{n(r)})^{d(r)} x^p$ where $p = n - d(1)n(1) - \ldots - d(r)n(r)$ is a non-negative integer since $d(1)n(1) + \ldots + d(r)n(r) = h(\omega) \leq n$. QED

Theorem 16.6 [45; Proposition 9.5, p. 256]. *Let $f: X \to Y$ be a flat morphism of finite type of Noetherian schemes. For any point $x \in X$, let $y = f(x)$. Then $\dim_x(X_y) = \dim_x X - \dim_y Y$.*

Corollary 16.7. *Suppose that A is a finitely generated integral domain over k. Then, $\dim A = \dim \mathrm{gr}A = \dim R$.*

Proof. We have seen that the inclusion mapping $i: k[x] \to D$ is flat and that Spec A and Spec(gr$A$) are fibers of i (Theorem 15.14). Then dim A = dim gr$A$ by Theorem 16.6. Since R is integral over gr$A$ (by Theorem 15.5), dim R = dim gr$A$ (by [71; Exercise 9.2, p.69]). QED

B. Modules

Theorem 16.2 has a number of powerful and (almost) immediate consequences to which we now turn. The first two apply to modules via a standard trick. For a moment, let A be any commutative k-algebra and let M be an A-module. We make $A \oplus M$ into a commutative k algebra by defining $(a + m)(a' + m'):= aa' + (am' + a'm)$. Then, one checks easily that $A \oplus M$ is finitely generated if and only if A is a finitely generated k-algebra and M is a finitely generated A-module.

Let A and M be as in the preceding paragraph and let G act rationally on both (and, as a group of algebra automorphisms on A). We say that these actions are *compatible* with the structure of M as an A-module if $g \cdot am = (g \cdot a)(g \cdot m)$ for all $g \in G$, $a \in A$, and $m \in M$.

Theorem 16.8. *Let A be a finitely generated, commutative k-algebra. Let M be an A-module. Let G be a reductive group which acts rationally on both A and M. Suppose that the actions of G are compatible with the structure of M as an A-module. Then M is a finitely generated A-module if and only if M^U is a finitely generated A^U-module.*

Proof. We form the k-algebra $A \oplus M$ as above and observe that G acts rationally on

$A \oplus M$. By Theorem 16.2, $A \oplus M$ is finitely generated if and only if $A^U \oplus M^U$ is. By assumption A is finitely generated, hence, so is A^U (Theorem 9.4). Thus, the theorem follows from the remarks preceding it. QED

Theorem 16.9. *Let A be a finitely generated, commutative k-algebra. Let M be an A-module. Let G be a reductive group which acts rationally on both A and M. Suppose that the actions of G are compatible with the structure of M as an A-module. If M is a finitely generated A-module, then M^G is a finitely generated A^G-module.*
Proof. (This argument follows immediately from Theorem A and does not require the machinery built up in this section.) If M is a finitely generated A-module, then the algebra $A \oplus M$ is finitely generated over k. Thus, by Theorem A, $(A \oplus M)^G = A^G \oplus M^G$ is finitely generated over k. But then M^G is a finitely generated A^G-module. QED

Theorem 16.10. *Let V be a rational G-module. Then V is finite-dimensional if and only if V^U is.*
Proof. We apply Theorem 16.8 to the case where $V: = M$, $k: = A$, and G acts trivially on k. QED

Theorem 16.11. *Let A be a finitely generated, commutative k-algebra on which the reductive group G acts rationally. Let C be a finitely generated, G-invariant subalgebra. Then A is integral over C if and only if A^U is integral over C^U.*
Proof. If A is integral over C, then A is a finite C-module. Then, by Theorem 16.8, A^U is a finite C^U-module and, so, A^U is integral over C^U. Conversely, if A^U is integral over C^U, then A^U is a finite C^U-module (using Theorem 9.4). Then A is a finite C-module by Theorem 16.8 and, so, A is integral over C. QED

Definition. Let A be a finitely generated, commutative k-algebra on which the reductive group G acts rationally. Let M be a finite-dimensional G-module. The *module of covariants of type M* is $B(M) = (A \otimes M)^G$.

Obviously, $A \otimes M$ is an A-module. The algebra A^G acts on $B(M) = (A \otimes M)^G$ by $a \cdot (a' \otimes m) = aa' \otimes m$. The basic fact about this action is

Lemma 16.12. *The A^G-module $B(M)$ is finitely generated over A^G.*
Proof. Since M is a finite-dimensional G-module, we see that $A \otimes M$ is a finitely generated A-module. Then, by Theorem 16.9, $(A \otimes M)^G$ is a finitely generated A^G-module. QED

- From now on, we shall assume that G is connected.

Definition. Let $\omega \in X^+(T)$ and let $B(\omega) = (A \otimes E(\iota\omega))^G$.

According to Theorem 12.4, the subspace $B(\omega)$ is isomorphic to $A(U,\omega) = \{a \in A : b \cdot a = \omega(b)a \text{ for all } b \in B\}$ via the mapping $\Sigma_i a_i \otimes f_i \to \Sigma_i f_i(e)a_i$. Furthermore, if char $k = 0$, then $\oplus_\omega(B(\omega) \otimes E(\omega))$ is actually isomorphic as a G-module to A (Theorem 12.8).

Lemma 16.13 [14; Corollaire, p.3]. *Let A be a finitely generated, commutative G-algebra. The following statements are equivalent.*
(a) *R is a flat A^G-module.*
(b) *R is a projective A^G-module.*
(c) *For any $\omega \in X^+(T)$, $B(\omega)$ is a projective A^G-module.*
(d) *For any $\omega \in X^+(T)$, $B(\omega)$ is a flat A^G-module.*

Proof. These statements are all quick consequences of standard facts about flat and projective modules and the fact that $R = \Sigma_\omega B(\omega) \otimes Y(\omega)$. First, a module of finite presentation is flat if and only if it is projective [71; Corollary, p.53]. Since A^G is finitely generated over k, any finite module is of finite presentation [71; Exercise 3.7, p.19]. This proves the equivalence of statements (c) and (d). We now prove the equivalence of statements (a) and (d). If a module is a direct sum, then it is flat if and only if each summand is flat [67; Proposition 3.1, p.613]. We need to show that $B(\omega)$ is flat if and only if $B(\omega) \otimes Y(\omega)$ is. Suppose that $B(\omega)$ is flat and let $\ldots \to M \to M' \to M'' \to \ldots$ be an exact sequence of $C = A^G$ modules. Since $B(\omega)$ is flat, the sequence $\ldots \to B(\omega) \otimes_C M \to B(\omega) \otimes_C M' \to B(\omega) \otimes_C M'' \to \ldots$ is also exact. Then, $\ldots \to (B(\omega) \otimes_C M) \otimes Y(\omega) \to (B(\omega) \otimes_C M') \otimes Y(\omega) \to (B(\omega) \otimes_C M'') \otimes Y(\omega) \to \ldots$ is exact. But (by [71; Formula 10, p.268]), $(B(\omega) \otimes_C M) \otimes Y(\omega) = (B(\omega) \otimes Y(\omega)) \otimes_C M$ for any A^G-module so we see that $B(\omega) \otimes_C Y(\omega)$ is flat. The converse is proved by reading this sequence of statements in reverse order.

Since projective modules are flat [55; Proposition 3.1, p.613], (b) implies (a). Finally, we prove that (c) implies (b). Indeed, since $B(\omega)$ is projective, so is $B(\omega) \otimes Y(\omega)$ and, then, R since it is a direct sum of projective modules [55; Proposition 3.5, p.193]. QED

Exercises.

1. Let H be a unipotent subgroup of G such that $k[G/H]$ is finitely generated over k. Show that A is finitely generated over k if and only if A^H is. (Hint: A is integral over $G{\cdot}A^H$.)
2. Suppose that char $k = 0$. Show that in the statement of Lemma 16.13, the algebra R may be replaced by A.

§17. S - varieties

• We assume in this section that G is connected.

Our goal in this section is to prove Theorem 17.4. Before getting to that, however, we need some preparatory lemmas.

Lemma 17.1. *Let H be an observable subgroup of G which contains the maximal unipotent subgroup U.*
(a) *H is normalized by the maximal torus T and there are elements $t_i \in T$ such that $H = \cup t_i H^\circ$.*
(b) *$k[G/H] = \oplus_\omega E(\omega)$ for certain ω, is finitely generated over k and is multiplicity-free.*
(c) *Let V be a finite-dimensional G-module and let $v \in V$ be such that $H = G_v$. Let v*

$= \Sigma_i v_i$ *where the* v_i *are non-zero T-weight vectors corresponding to distinct T-weights* ω_i. *Then, H fixes each* v_i.

Proof. First, let us suppose that H is connected. Let $B_H = T_1U$ be a Borel subgroup in H. The torus T_1 is contained in the normalizer in G of U which is TU (Exercise 1). Since the homomorphism $T_1 \rightarrow (T \cap H)^\circ$ given by $t_1 = tu \rightarrow t$ is injective, we see that $\dim T_1 = \dim (T \cap H)^\circ$. Thus, we may replace T_1 by $(T \cap H)^\circ$. It follows that T normalizes B_H. Now, there is a finite-dimensional G-module V and a vector $v \in V$ such that $H = G_v$. For any $b \in B_H$ and $t \in T$, we have $tbt^{-1}\cdot v = v$. But, tB_Ht^{-1} is a Borel subgroup in tHt^{-1} so $tht^{-1}\cdot v = v$ for all $t \in T$ and $h \in H$. Thus, $tht^{-1} \in H$.

Next, suppose that H is not connected. The subgroup H° is observable (Corollary 2.2) and, by what was just proved, is normalized by T. Consider a coset hH° of H° in H. The subgroup hUh^{-1} is maximal unipotent in H° and, so, there is an $h_1 \in H^\circ$ such that $hUh^{-1} = h_1Uh_1^{-1}$. Therefore, $h_1^{-1}h$ is in the normalizer in G of U which is TU (Exercise 1). Thus, the coset $hH^\circ = tH^\circ$ for some $t \in T$ and, so, H is normalized by T.

To prove statement (b), we first note that $k[G/H] \subset k[G/U] = \oplus_\omega E(\omega)$. As we just showed, there is a subgroup $T_1 = (T \cap H)$ of H such that T_1U is epimorphic in H, i.e., H fixes (under right translation) a function $f \in k[G]$ if and only if T_1U does. It follows that $k[G/H] = \oplus_\omega E(\omega)$ where $\omega \mid T_1 = 1$. The algebra $k[G/H]$ is finitely generated over k by Corollary 9.5 and is multiplicity-free according to Example (2), §11, Section A. Finally, statement (c) follows by a small modification of Lemma 3.8. QED

Definition. A subgroup of G which contains a maximal unipotent subgroup of G is called *horospherical*. Let X be an affine variety on which G acts morphically. Then, X is said to be an *S-variety* if there is an element $x \in X$ such that (i) the orbit $G{\cdot}x$ is open, dense in X and (ii) G_x is horospherical.

Lemma 17.2. *Let X be an S-variety; let* $x \in X$ *be such that the orbit* $G{\cdot}x$ *is open, dense in X and* $H = G_x$ *contains U. Let* $R = \oplus_\omega E(\omega)$ *where the sum is over all* ω *corresponding to non-zero T-weight functions in* ${}^Uk[X]$. *Then R is a finitely generated k-algebra invariant under right translation by T and* ${}^UR$ *is T-isomorphic to* ${}^Uk[X]$. *Let Y be the affine variety corresponding to R. Then, Y is an S-variety and there is a point* $y \in Y$ *such that* $G{\cdot}y$ *is open in Y and* $(G_y)^\circ = H^\circ$. *Furthermore,* $\dim(X - G{\cdot}x) \leq \dim X - 2$ *if and only if* $\dim (Y - G{\cdot}y) \leq \dim Y - 2$.

Proof. Let ${}^Uk[X] = k[f_1, \ldots, f_r]$ where f_i is a T-weight vector having T-weight ω_i. Let $\varphi\colon G \rightarrow X$ be given by $\varphi(g) = g{\cdot}x$ and let φ^* be its comorphism (see Exercise 3, §1). Now, φ^*f_i is fixed under left translation by U and has left T-weight ω_i. According to Lemma 17.1(b), φ^*f_i must be the unique vector in $E(\omega_i)$ which is fixed by U. Let $R = \oplus_\omega E(\omega)$ where the sum is over all ω of the form $c_1\omega_1 + \ldots + c_r\omega_r$, $c_i \in \mathbf{Z}$, $c_i \geq 0$. Then, R is a finitely generated k-algebra by Theorem 16.2 since ${}^UR = k[f_1, \ldots, f_r]$; also, R is invariant under left translation by G. Let S be the finitely generated k-algebra generated by all the $\langle G{\cdot}f_i \rangle$. Then, $S \subset k[X]$ and ${}^US = {}^Uk[X]$ since $k[f_1, \ldots, f_r] \subset {}^US \subset {}^Uk[X] = k[f_1, \ldots, f_r]$. It follows from Theorem 16.11 that $k[X]$ is integral over S. Similarly, we see that ${}^UR = {}^US$ so that R is integral over S. In particular, $\dim k[X] = \dim R = \dim S$.

Let Y be the affine variety such that $k[Y] = R$ and let Z be the affine variety such that $k[Z] = S$. Then, G acts on Y and Z. Let $\pi: X \to Z$ and $\pi': Y \to Z$ be the maps corresponding to the inclusions $k[Z] \subset k[X]$ and $k[Z] \subset k[Y]$. Suppose that $\pi(x) = z$ and $\pi'(y) = z$. Since both π and π' are G-equivariant, we see that $H = G_x \subset G_z$ and $G_y \subset G_z$. Since $\dim Y = \dim X = \dim Z$, we see that $\dim G_y = \dim H$. The last statement in the lemma follows (as in the first paragraph of the proof of Theorem 4.3) since both $k[Y]$ and $k[X]$ are integral extensions of $k[Z]$ and dimension is preserved under integral extensions. QED

For a moment, let X be a variety on which G acts morphically and suppose that there is an element $v \in X$ such that the orbit $G{\cdot}v$ is open and dense in X. Let $\varphi: G \to X$ be the mapping given by $\varphi(g) = g{\cdot}v$ and let φ^* be the comorphism of φ. Then, $\varphi^*(f)(g) = f(g{\cdot}v)$ for all $f \in k[X]$ and the algebra $k[X]$ can be identified with a subalgebra, $\varphi^*(k[X])$, of $k[G]$ (Exercise 3, §1). We shall say that $k[X]$ is *invariant under right translation by* T if the subalgebra $\varphi^*(k[X])$ is.

Lemma 17.3. *Let X be an S-variety, contained in a finite-dimensional G-module V, and suppose that the orbit $G{\cdot}v$ is open, dense in X where G_v contains U. Let $v = v_1 + \ldots + v_s$ where the v_i are non-zero T-weight vectors having distinct T-weights ω_i. There is an S-variety Y, contained in a finite-dimensional G-module W, and a point $w \in Y$ such that (i) the orbit $G{\cdot}w$ is open in Y and G_w contains U; (ii) $w = w_1 + \ldots + w_s$ where the w_i are non-zero T-weight vectors having distinct T-weights ω_i; (iii) the algebra k[Y] is invariant under right translation by T; (iv) $\dim(X - G{\cdot}v) \leq \dim X - 2$ if and only if $\dim(Y - G{\cdot}w) \leq \dim Y - 2$.*

Proof. We shall apply Theorem 12.4 and its proof. For a given $i = 1, \ldots, s$, the vector v_i is in $V(U,\omega_i)$. By Theorem 12.4, there is a linear, G-equivariant mapping ψ_i: $E(\iota\omega_i)^* \to V$ such that $\psi_i(w_i) = v_i$ where w_i is the unique (up to scalar) non-zero element in $E(\iota\omega_i)^*$ which is fixed by U and has T-weight ω_i. Let $W = E(\iota\omega_1)^* \oplus \ldots \oplus E(\iota\omega_s)^*$ and let $w = w_1 \oplus \ldots \oplus w_s$. Let $\varphi: W \to V$ be defined by $\varphi(z_1 \oplus \ldots \oplus z_s) = \psi_1(z_1) + \ldots + \psi_s(z_s)$. Then, φ is G-equivariant and $\varphi(w) = v$. Let $Y = \mathrm{cl}(G{\cdot}w)$. We now show that Y has the desired properties.

First, Y is an S-variety since each w_i is fixed by U. According to Lemma 17.1(c), $\dim Y = \dim X$. Also, $k[Y]$ is invariant under right translation by T. Indeed, let $\mu \in W^*$. By our definition of W, we may assume that μ is in one of the $(E(\iota\omega_i)^*)^*$. Let $p: W \to E(\iota\omega_i)^*$ be the projection mapping; then, p is G-equivariant. To show that $k[Y]$ is invariant under right translation by T, we need only show that $\mu(gt{\cdot}w) = \omega_i(t)\mu(g{\cdot}w)$ for all $g \in G$ and $t \in T$. But, $\mu(gt{\cdot}w) = \mu(p(gt{\cdot}w)) = \mu(gtw_i) = \omega_i(t)\mu(g{\cdot}w)$.

Next, we study the closures of the two orbits $G{\cdot}v$ and $G{\cdot}w$. Now, $\mathrm{cl}(G{\cdot}v) = G{\cdot}\mathrm{cl}(B{\cdot}v)$ by Theorem 5.3. But U fixes v so $B{\cdot}v = T{\cdot}v$ where $t{\cdot}v = \omega_1(t)v_1 + \ldots + \omega_s(t)v_s$. It follows that an element in the closure of $B{\cdot}v$ has the form $c_1v_1 + \ldots + c_sv_s$ where some of the c_i's are zero. Suppose that this element is $c_1v_1 + \ldots + c_pv_p$ where $p < s$ and each of the c_i's is non-zero. The stabilizer in G of this element is generated by $\{t \in T : \omega_i(t) = 1 \text{ for } i = 1, \ldots, p\}$ and $\{U_\alpha : (\omega_i,\alpha) \geq 0\}$, where we have used Theorem 3.2 and Lemma 17.1(c). Exactly the same sequence of statements holds for $\mathrm{cl}(G{\cdot}w)$. So, the finitely many G-orbits in each closure (see Exercise 3) have precisely the same form. Thus, for example, suppose that an orbit in $X - G{\cdot}v$ has codimension 1. Then, the corresponding orbit in $Y - G{\cdot}w$ also has codimension 1. QED

Theorem 17.4. *Let X be an S-variety (contained in a finite-dimensional G-module V) and suppose that the orbit* $G{\cdot}v$ *is open in X where* G_v *contains U. Let* $v = \Sigma_i v_i$ *where the* v_i *are non-zero T-weight vectors having distinct T-weights* ω_i. *Let* $\mathbf{Z}E = \{\Sigma_i c_i\omega_i : c_i \in \mathbf{Z}\}$ *and* $\mathbf{Q}_+E = \{\Sigma_i c_i\omega_i : c_i \in \mathbf{Q},\ c_i \geq 0\}$. *Then* $\dim(X - G{\cdot}v) \leq \dim X - 2$ *if and only if* $\mathbf{Z}E \cap X^+(T) \subset \mathbf{Q}_+E$.

Proof. According to Lemma 17.3, we may assume that $k[X]$ is invariant under right translation by T. Let $H = G_v$. Let $\varphi: G \to X$ be given by $\varphi(g) = g{\cdot}v$ and let φ^* be its comorphism. Let $f \in \varphi^*(k[X])$ and have right T-weight μ. We first show that $\mu \in \mathbf{Z}_+E = \{\Sigma_i c_i\omega_i : c_i \in \mathbf{Z},\ c_i \geq 0\}$. Indeed, for $\lambda \in V^*$, we have $\lambda(g{\cdot}v) = \Sigma_i \lambda(g{\cdot}v_i)$. But each of the $f_i(g) = \lambda(g{\cdot}v_i)$ is in $k[G/H]$ (by Lemma 17.1(c)) and f_i has right T-weight ω_i. Since $\varphi^*(k[X])$ is invariant under right translation by T and is generated by all the $\lambda(g{\cdot}v)$, we see that $\mu \in \mathbf{Z}_+E$. The same argument shows that if $\eta \in \mathbf{Z}_+E$, then there is an $F \in k[X]$ having η as right T-weight.

Now, suppose that $\dim(X - G{\cdot}v) \leq \dim X - 2$. Let $S = T \cap H$ and let $\omega \in \mathbf{Z}E \cap X^+(T)$. Since H fixes each v_i, each $\omega_i | S$ is trivial. But, $\omega \in \mathbf{Z}E$ so $\omega | S$ is also trivial. Next, let $\alpha \in \Phi(T,H)$; we shall show that $(\omega,\alpha) \geq 0$. If $\alpha \in \Phi^+$, this holds since $\omega \in X^+T$. If $-\alpha \in \Phi^+ \cap \Phi(T,H)$, then each $(\omega_i, \alpha) = 0$ since H fixes v_i (Theorem 3.2). Thus, $(\omega, \alpha) = 0$. Let $f \in k[G/H]$ be non-zero and have right T-weight ω (Corollary 3.7, slightly modified). Since $\operatorname{codim}(X - G{\cdot}v) > 1$, $k[G/H]$ is integral over $k[X]$ by the proof of Theorem 4.3. Hence, f satisfies an equation of the form

$$f^m + a_1 f^{m-1} + \ldots + a_m = 0$$

where $a_i \in k[X]$ and $a_m \neq 0$. We may assume that each a_i is a right T-weight vector (since f is) and, so, is in $\mathbf{Z}_+E$ by the argument given in the first paragraph of this proof. If a_m has weight η, then $m\omega = \eta$ and $\omega \in \mathbf{Q}_+E$.

Now, assume that $\mathbf{Z}E \cap X^+(T) \subset \mathbf{Q}_+E$. Let $f \in {}^Uk[G/H]$. We will show that f is integral over ${}^Uk[X]$. Then, $k[G/H]$ is integral over $k[X]$ by Theorem 16.11 and $\dim(X - G{\cdot}v) \leq \dim X - 2$ by Theorem 4.3. We may assume that f has right T-weight ω. By Theorem 12.1(b), f has left T-weight $\iota\omega$ and is the only function in ${}^Uk[G/H]$ having this weight since $k[G/H]$ is multiplicity-free. Now, $\omega | S$ is trivial so there is a positive integer m satisfying $m\omega \in \mathbf{Z}E$ (Exercise 2). Therefore, $m\omega \in \mathbf{Z}E \cap X^+T$ and, by assumption, there is an $\eta \in \mathbf{Z}_+E$ so that $m\omega = (m_1/n_1)\eta$ for certain integers m_1 and n_1. Since $\eta \in \mathbf{Z}_+E$, there is an $F \in {}^Uk[X]$ having right T-weight η. Then, we see that

$$f^{n_1 m} = cF^{m_1}$$

by comparing right T-weights, always bearing in mind that $k[G/H]$ is multiplicity-free. QED

Theorem 17.5. *Let X be an S-variety and suppose that the orbit* $G{\cdot}v$ *is open, dense in X. Suppose that* ${}^Uk[X] = k[f_1, \ldots, f_r]$ *where* f_i *is non-zero and has T-weight* ω_i. *Let* $\mathbf{Z}E = \{\Sigma_i c_i\omega_i : c_i \in \mathbf{Z}\}$ *and* $\mathbf{Q}_+E = \{\Sigma_i c_i\omega_i : c_i \in \mathbf{Q},\ c_i \geq 0\}$. *Then* $\dim(X - G{\cdot}v) \leq \dim X - 2$ *if and only if* $\mathbf{Z}E \cap X^+(T) \subset \mathbf{Q}_+E$.

Proof. Let Y and $y \in Y$ be as in Lemma 17.2; we change notation and suppose that $y = y_1 + \ldots + y_s$ where the y_i are non-zero T-weight vectors having distinct T-weights η_i.

We showed in the first paragraph of the proof of Theorem 17.4 that the right T-weights on $k[Y]$ are $\mathbf{Z}_+E' = \{c_1\eta_1 + \dots + c_s\eta_s : c_i \in \mathbf{Z}, c_i \geq 0\}$. The subspace of $k[Y]$ consisting of those functions having right T-weight η is invariant under left translation by G and has a U-fixed point. According to Lemma 17.1, the left T-weight of any such non-zero U-fixed point must be $\iota\eta$. Thus, the left T-weights on ${}^Uk[Y]$ are $\mathbf{Z}_+E'' = \{c_1\iota\eta_1 + \dots + c_s\iota\eta_s : c_i \in \mathbf{Z}, c_i \geq 0\}$. But, this set is also $\mathbf{Z}_+E$. Applying Theorem 17.4, we see that the following statements are equivalent: (i) $\dim(X - G{\cdot}v) \leq \dim X - 2$; (ii) $\dim(Y - G{\cdot}y) \leq \dim Y - 2$; (iii) $\mathbf{Z}E' \cap X^+(T) \subset \mathbf{Q}_+E'$; (iv) $\mathbf{Z}E'' \cap X^+(T) \subset \mathbf{Q}_+E''$; (v) $\mathbf{Z}E \cap X^+(T) \subset \mathbf{Q}_+E$. QED

Exercises.

1. Show that the normalizer in G of U is $B = TU$.

2. Let $\chi_i \in X(T)$, $1 \leq i \leq r$, and let $S = \{t \in T : \chi_i(t) = 1 \text{ for all } i = 1, \dots, r\}$. Let $\chi \in X(T)$ be such that $\chi \mid S = 1$. Show that there is an integer m such that $m\chi = c_1\chi_1 + \dots + c_r\chi_r$ for certain integers c_i.

3. Prove the statement in the last paragraph of the proof of Lemma 17.3 concerning finitely many T-orbits. (Hint: Let $v_1, \dots, v_s$ be T-weight vectors having distinct T-weights ω_i and let $v = v_1 + \dots + v_s$. We describe the (finitely many) orbits in the closure of $T{\cdot}v$. If the ω_i are linearly independent over $\mathbf{Q}$, this is immediate. Otherwise suppose that $\omega_1, \dots, \omega_r$ is a maximal linearly independent subset. Then, the action of T on v_p, $p \geq r$, is completely described by the action on $v_1, \dots, v_r$ up to a root of unity.)

§18. Flat Deformations and Normality

- We shall assume in this section that G is connected.

In this section, we shall relate various ring-theoretic properties of A to those of A^U. We begin by studying nilpotent elements and zero divisors and then turn to normality.

Definition. Let M be an A-module. A prime P in A is called an *associated prime* of M if there is an element $x \in M$ such that $P = \{a \in A : ax = 0\}$.

Lemma 18.1 [71; 6.1, 6.5, pp.38,39]. *Let B be a Noetherian ring and let M be a non-zero, finitely generated B-module. Then (i) the set of associated primes of M is finite and (ii) the set of zero divisors of M is the union of all the associated primes of M.*

Theorem 18.2. *The algebra A has nilpotent elements (resp. zero divisors) if and only if A^U does.*

Proof. The "if" direction in both cases is immediate. Now, suppose that A has nilpotent elements and let V be the algebra consisting of all the nilpotent elements in A. Then V is invariant under G so $V^U \neq \{0\}$. Next, suppose that A has zero divisors; we may assume that A is finitely generated over k. Let $Z(A)$ be the set of all zero divisors in A. Let $P_1,\dots,P_n$ be the associated primes of A so (by Lemma 18.1), $Z(A) = P_1 \cup \dots \cup P_n$. The group G acts as a permutation group on $\{P_1,\dots,P_n\}$. Since G is connected, this action is trivial and P_1, say, is G-invariant. Let $b \in P_1^U$, $b \neq 0$. Let $V = \{a \in A : ba = 0\}$. Then V is a non-zero vector space which is invariant under U. Thus, $V^U \neq \{0\}$. QED

Next, we study "normality." We recall that a ring A is said to be *normal* if the localization of A at any prime P is an integrally closed domain. To study the relationship between the normality of A and the normality of A^U, we need a variety of facts.

(R1) [71; Remark, p.64] *If A is a normal, Noetherian ring and $P_1,\ldots,P_r$ are all the minimal primes of A, then $A \simeq A/P_1 \times \ldots \times A/P_r$ and each A/P_i is an integrally closed domain. Conversely, the direct product of a finite number of integrally closed domains is normal.*

(R2) [73, p.278] A point $P \in$ Spec A is called *normal* if A_P is an integrally closed domain. The set of normal points is open in Spec A.

(R3) [73; Lemma 2, p.126] *Let X be a prescheme of finite type over k. Then the closed points of X are dense in every closed subset of X.*

(R4) [71; Corollary, p.184] *Let (B_1,M_1) and (B_2,M_2) be Noetherian local rings and let $f: B_1 \to B_2$ be a local homomorphism. Suppose that B_2 is flat over B_1. If both B_1 and the fiber rings of f are normal, then so is B_2.*

(R5) [71; Theorem 7.1, p.46] *Let $f: B \to C$ be a ring homomorphism. A necessary and sufficient condition for C to be flat over B is that for every prime ideal Q of C, C_Q is flat over B_P where $P = f^{-1}(Q)$.*

Theorem 18.3. *Suppose that A is finitely generated over k. If grA is normal, then A is normal.*

Proof. Suppose that A is not normal. Then we shall show that D is not normal. Indeed, if D were normal, then $D[1/x] = A[x,1/x]$ would be normal by Lemma 15.12. Now let P be any prime ideal in A. Then $PA[x]$ is a prime ideal in $A[x]$ which does not contain x. Using Lemma 15.12 again, we see that $A[x]_{PA[x]}$ is an integrally closed domain. But, then, A_P is integrally closed. For let f be an element in the quotient field of A_P which is integral over A_P. We may consider A_P to be a subdomain of $A[x]_{PA[x]}$. Since the latter ring is integrally closed, $f \in A[x]_{PA[x]}$. Let $f(x) = r(x)/s(x) = a/b$ where $r(x)$, $s(x)$ are in $A[x]$, $s(x)$ is not in $PA[x]$, and $a,b \in A_P$. There is a polynomial $q(x)$ which is not in $PA[x]$ such that $bq(x)r(x) = aq(x)s(x)$. Now $q(x)s(x) \notin PA[x]$; suppose that the coefficient d of x^m in $q(x)s(x)$ is not in P. Then, comparing coefficients of x^m on both sides of the preceding equation, we see that there are elements $c,d \in A$ so that $bc = ad$. Then $f = a/b = c/d$ is in A_P. Thus, A_P is integrally closed, contradicting the fact A is not normal. So, from now on, we may assume that D is not normal.

The set of points $P \in$ Spec D such that D_P is not normal is a closed non-empty subset D' of D by (R2). The group k^* acts as a group of automorphisms of D and D' is (obviously) invariant under k^*. Furthermore, D' contains a maximal ideal M by (R3). Suppose that $M \cap k[x] = (x - \alpha)$, the principal ideal in $k[x]$ generated by $x - \alpha$, where

$\alpha \neq 0$. We now show that there is a maximal ideal M' in D' so that $M' \cap k[x] = (x)$. Indeed, since $x \notin M$, the ideal $MD[1/x] = MA[x,1/x]$ is proper and maximal. Let $N = MA[x,1/x] \cap A[x]$. Then $N \cap D$ contains M and, so, $N \cap D = M$. The inclusion of D in $A[x]$ gives rise to a continuous, k^*-equivariant mapping φ: Spec $A[x] \to$ Spec D such that $\varphi(N) = M$. The closed, k^*-invariant set $\varphi^{-1}(D')$ contains N. According to Lemma 15.13, there is a maximal ideal N' in $\varphi^{-1}(D')$ which contains x. Let M' be any maximal ideal in D which contains $N' \cap D$.

Let us change notation and write M instead of M'. We now show that D_M is normal. Since $M \in D'$, this will be the desired contradiction. Let $P = M \cap k[x] = (x)$. Since D is flat over $k[x]$ (by Lemma 15.9), we see that D_M is flat over $k[x]_P$ by (R5). According to (R4), we need only show now that the fiber rings are normal. But the only fiber ring is over the ideal $Pk[x]_P$ and this ring is D_M/PD_M by Lemma 15.10. We now apply Lemma 15.11 to see that D_M/PD_M is a localization of D/xD; since D/xD is isomorphic to gr$A$ (by Lemma 15.8), it is normal. QED

Theorem 18.4. *If A is normal, then A^U is normal. If A^U is normal, if A is finitely generated over k, and if gr$A = R$, then A is normal.*
Proof. Let A be normal; we shall show that A^U is normal. By (R1), we may assume that A is an integral domain. Let a be an element in the quotient field Q of A^U which is integral over A^U. Then a is integral over A, so, $a \in A \cap Q = A^U$. Next, suppose that A^U is normal and gr$A = R$. Then $R = (A^U \otimes k[G/U^-])^T$ is normal. (In general, if B is a rational, T-algebra which is normal, then so is B^T. For let b be an element in the quotient field Q of B^T which is integral over B^T. Then $b \in B \cap Q = B^T$.) Hence, gr$A$ is normal and we may apply Theorem 18.3. QED

As mentioned in the note following Theorem 15.5, if char $k = 0$, then the condition gr$A = R$ is automatically satisfied since we have seen that $(\text{gr}A)^U = R^U$. If char $k \geq 0$, it may be shown that A has a good filtration if and only if gr$A = R$ [42; Theorem 16, p.133]. Theorem 18.4 was proved in characteristic 0 in [113; Theoreme 1, p.8].

Example: an algebra A such that A^U is normal but A is not. We continue the example which concluded §14 but now take char $k = 3$. The algebra $A = S(V)$ is integrally closed, hence, so is A^U according to Theorem 18.4. We note that $A^U \supseteq (G{\cdot}A^U)^U \supseteq A^U$ so $(G{\cdot}A^U)^U = A^U$. We now show that e_2 is in the quotient field of $G{\cdot}A^U$. The verification is by direct calculation; we determine elements in A^U and then apply $u \in U^-$ to find elements in $G{\cdot}A^U$. (i) $e_0 \in A^U \Rightarrow e_3 \in G{\cdot}A^U$. (ii) $e_0e_2 - e_1^2 \in A^U \Rightarrow e_0e_3 - e_1e_2$ and $e_1e_3 - e_2^2$ are in $G{\cdot}A^U$. (iii) $e_1e_2 \in G{\cdot}A^U$ by (i) and (ii). (iv) The element e_2 is integral over $G{\cdot}A^U$ since $e_2^3 + e_2(e_1e_3 - e_2^2) - e_1e_2e_3 = 0$. (v) $2e_0^2e_3 + e_1^3 \in A^U \Rightarrow e_0e_3^2 - e_2^3 \in G{\cdot}A^U \Rightarrow e_2^3 \in G{\cdot}A^U \Rightarrow e_2$ is in the quotient field of $G{\cdot}A^U$ by (iv). Therefore, e_2 is integral over $G{\cdot}A^U$ and is in the quotient field of $G{\cdot}A^U$; but, as we saw earlier $e_2 \notin G{\cdot}A^U$. Thus, $G{\cdot}A^U$ is not normal.

Example: determinantal varieties. Let $M_{m,n}$ be the space consisting of all $m \times n$ matrices over k. Let $D_{m,n,r}$ be the subvariety of $M_{m,n}$ consisting of all matrices having rank $\leq r$; $D_{m,n,r}$ is called a *determinantal variety*. The group $G = GL_m \times GL_n$ acts on $M_{m,n}$ and on $D_{m,n,r}$ by $(g,h){\cdot}x = gxh^{-1}$.

Lemma 18.5. *$\dim D_{m,n,r} = r(m + n - r)$.*

Proof. Using facts from elementary linear algebra, we see that the set of matrices having rank r is the G-orbit of $x_o =$

$$\begin{pmatrix} I_r & 0_{r,n-r} \\ 0_{m-r,r} & 0_{m-r,n-r} \end{pmatrix}.$$

The stabilizer, H, of x_o consists of all matrices of the form

$$\begin{pmatrix} A_{r,r} & B_{r,m-r} \\ 0_{m-r,r} & C_{m-r,m-r} \end{pmatrix} \times \begin{pmatrix} A_{r,r} & 0_{r,n-r} \\ D_{n-r,r} & E_{n-r,n-r} \end{pmatrix}.$$

Then, $\dim D_{m,n,r} = \dim G - \dim H = m^2 + n^2 - [r^2 + m(m - r) + n(n - r)] = r(m + n - r)$. QED

Corollary 18.6. *Let H be as above. Then, H is a spherical subgroup of G.*

Proof. Let B^- (resp. B^+) be the Borel subgroup of GL_m (resp. GL_n) consisting of all lower (resp. upper) triangular matrices. The $B^- \times B^+$-orbit of x_o is open in $D_{m,n,r}$ since $\dim (B^- \times B^+) - \dim ((B^- \times B^+) \cap H) = \dim D_{m,n,r}$. Therefore, H is a spherical subgroup of G according to Theorem 11.1. QED

Let U^- (resp. U^+) be the maximal unipotent subgroup of GL_m (resp. GL_n) consisting of all lower (resp. upper) triangular matrices with 1's on the diagonal and let $U = U^- \times U^+$. We next calculate the algebra of U-invariants on $D_{m,n,r}$ under the assumption that $m \leq n$. To that end, we define functions D_i on $M_{m,n}$ as follows: for $x \in M_{m,n}$, let $D_i(x)$, $1 \leq i \leq m$, be the determinant of the matrix obtained from x by choosing its first i rows and columns. According to Exercise 1, §13, the D_i are U-invariant and, in fact, $k[M_{m,n}]^U = k[D_1, \dots, D_m]$. In what follows, we shall denote the restrictions of the D_i to $D_{m,n,r}$ by D_i, also.

Lemma 18.7. *The algebra of U-invariants on $D_{m,n,r}$ is $k[D_1, \dots, D_r]$, a polynomial algebra.*

Proof. The restriction mapping from $k[M_{m,n}]$ to $k[D_{m,n,r}]$ sends $k[M_{m,n}]^U$ to $k[D_{m,n,r}]^U$ and, in fact, $k[D_{m,n,r}]^U$ is integral over the image of $k[M_{m,n}]^U = k[D_1, \dots, D_r]$ by Theorem 9.9. Let X be the subset of $D_{m,n,r}$ consisting of all matrices of the form

$$\begin{pmatrix} D_{r,r} & 0_{r,n-r} \\ 0_{m-r,r} & 0_{m-r,n-r} \end{pmatrix}$$

where $D_{r,r}$ is any diagonal matrix having non-zero diagonal entries. Now, $D_i | X$ is just the product of the first i diagonal elements in $D_{r,r}$, so the algebra $k[D_1, \dots, D_r]$ is a polynomial algebra and is integrally closed in its quotient field. Since the $B^- \times B^+$-orbit of x_o is open in $D_{m,n,r}$, it follows that the set $U \cdot X$ contains an open set in $D_{m,n,r}$. So, if f is any U-invariant function on $D_{m,n,r}$, then f is a rational function in the D_i. Hence, any function in $k[D_{m,n,r}]^U$ is actually in $k[D_1, \dots, D_r]$. QED

Corollary 18.8. *Let char $k = 0$. The variety $D_{m,n,r}$ is normal.*

Proof. According to Theorem 18.4, it is enough to prove that $k[D_{m,n,r}]^U$ is normal. But, we have just seen that this is a polynomial algebra. QED

Note. Corollary 18.8 holds in arbitrary characteristic. Indeed, $D_{m,n,r}$ has a good filtration [32; Proposition 1.4c, p.125]. Then, as noted above, grA = R and Theorem 18.4 again applies.

Exercise.

1 [112; Theorem 10, p.755]. Let X be an S-variety (contained in a finite-dimensional G-module V) and suppose that the orbit $G{\cdot}v$ is open in X where G_v contains U. Let $v = \Sigma_i v_i$ where the v_i are non-zero T-weight vectors having distinct T-weights ω_i. Let $\mathbf{Z}E = \{\Sigma_i c_i\omega_i : c_i \in \mathbf{Z}\}$ and $\mathbf{Q}_+E = \{\Sigma_i c_i\omega_i : c_i \in \mathbf{Q},\ c_i \geq 0\}$. If char $k = 0$, there is no further assumption. If char $k > 0$, we shall assume that $k[X]$ has a good filtration, i.e., $X = Y$ as in Lemma 17.2. Show that X is normal if and only if $\mathbf{Z}E \cap \mathbf{Q}_+E = \mathbf{Z}_+E$.

Bibliographical Note.

This chapter is based on [89] which assumes that k is of characteristic 0 and its extension to arbitrary characteristic in [42]. It should be noted that [89] has many interesting references. The material in §17 was first proved for char $k = 0$ in [112].

Chapter Four

Complexity

Introduction. The concept of the "complexity" of a group action was introduced by D. Luna and T. Vust in their work on equivariant embeddings of homogenous spaces; if G acts on a variety X and if $B = TU$ is a Borel subgroup of G, then the complexity of the action, $c(X)$, is the codimension in X of a B-orbit having highest dimension (Lemma 19.7(a)). In this chapter, we shall see how this concept arises in a variety of situations. In §19, the pertinent definitions are given and some basic facts are proved. In §20, $k[X]^U$ is shown to be finitely generated over k when $k[X]$ is a unique factorization domain and $c(X)$ is either 0 or 1; in addition, some results about the quotient space X/U and, also, multiplicities are proven. F. Knop has shown that the finite generation of $k[X]^U$ in case $c(X) = 0$ or 1 is true without the assumption of unique factorization and this is proved in §21. A closed subgroup H of G is said to be spherical if $c(G/H) = 0$. Such subgroups have arisen in several examples which we have already encountered; in §22, we give a more systematic approach to this subject. Finally, in §23, we return to the study of induced modules, considering the question: if $k[G]^H$ is finitely generated over k and if E is a finite-dimensional H-module, is $\mathrm{ind}_H{}^G E$ a finite $k[G]^H$-module?

§19. Basic Principles

Before coming to our main results, we state some lemmas which will be of value in several situations.

Lemma 19.1. *Let $H = TU$ be a connected solvable group, where T is a maximal torus in H and U is the subgroup consisting of all unipotent elements in H. Let A be an integral domain, with quotient field $Q(A)$, on which H acts rationally. Let $\chi \in X(T)$; let $f \in Q(A)$ satisfy $h{\cdot}f = \chi(h)f$ for all $h \in H$. Then, there are elements $a,b \in A^U$ and a character $\mu \in X(T)$ such that (i) $t{\cdot}a = (\mu + \chi)(t)a$ and $t{\cdot}b = \mu(t)b$ for all $t \in T$; (ii) $f = a/b$.*

Proof. Let $V = \{b \in A : bf \in A\}$. The vector space V is H-invariant and is non-zero. Since H acts rationally on A, there is a finite-dimensional, non-zero H-invariant subspace V_1 of V. Let b be a non-zero eigenvector of H in V_1; then $b \in A^U$ and there is a character $\mu \in X(T)$ such that $t{\cdot}b = \mu(t)b$ for all $t \in T$. Let $a = bf$; then a,b, and μ satisfy the desired conditions. QED

Corollary 19.2. *Let H be a unipotent group which acts rationally on an integral domain A. Let $f \in Q(A)^H$. Then, there are elements $a,b \in A^H$ such that $f = a/b$.*

Lemma 19.3. *Let X be an irreducible variety on which G acts morphically. For each non-negative integer r, $\{x \in X : \dim G{\cdot}x \leq r\}$ is closed in X. If $m = \max \{\dim G{\cdot}x : x \in X\}$, then the set $X' = \{x \in X : \dim G{\cdot}x = m\}$ is open and dense in X.*

Proof. Consider the mapping $G \times X \to X \times X$ defined by $(g,x) \to (g{\cdot}x,x)$. The fiber above a point $(g\cdot x,x)$ is (gG_x, x) and, so, has dimension equal to dim G_x. According to the principle of upper semi-continuity for fibers [10; Corollary 10.3, p.21], we see that $\{x \in X : \dim G_x \geq r\}$ is closed for each integer r. Thus, $\{x \in X : \dim G\cdot x = \dim G - \dim G_x \leq r\}$ is closed in X. In particular, $\{x \in X : \dim G\cdot x = \dim G - \dim G_x \leq m - 1\}$ is closed in X. QED

In what follows, we shall use the notion of a quotient morphism as defined in [10; §6]. We recall it here along with some basic facts.

Definition. Let V be a variety and suppose that G acts morphically on V. A morphism $\pi: V \to W$ is said to be a *quotient morphism* if (1) it is surjective and open; (2) its fibers are the orbits of G acting on V; (3) if $U \subset V$ is open and G-invariant, then the comorphism π^o induces an isomorphism from $k[\pi(U)]$ onto $k[U]^G$.

Lemma 19.4 [10; Propositions 6.4 and 6.5; pp.96-97]. *Let V be an irreducible variety and suppose that G acts morphically on V. Let $\pi: V \to W$ be a quotient morphism.*
(a) *The orbits of G in V have constant dimension $d = \dim V - \dim W$.*
(b) *π is a separable morphism and its comorphism induces an isomorphism of $k(W)$ onto $k(V)^G$.*

The next theorem is a famous result of M. Rosenlicht. We postpone its proof to the appendix.

Theorem 19.5. *Let G be a connected, affine algebraic group. Let X be an irreducible variety on which G acts morphically. Then, there is a dense, open subset X' of X such that the quotient of X' by G exists.*

Corollary 19.6. *Let X, X', and G be as in Theorem 19.5.*
(a) *Each orbit of G on X' is closed in X'.*
(b) *Let $\pi : X' \to Y$ be the quotient mapping. Then $k(Y) = k(X')^G = k(X)^G$.*
(c) *The orbits of G on X' have constant dimension $d = \dim X - \dim Y = \dim X -$ transc. deg. $k(X)^G$.*
(d) *Let $m = \max\{\dim G\cdot x : x \in X\}$. Then $m = d$.*
(e) *There is an $x \in X$ such that $G\cdot x$ is open in X if and only if $k(X)^G = k$.*
Proof. Statement (a) holds since each fiber of the quotient map is an orbit. Statements (b) and (c) follow at once from Lemma 19.4. Statement (d) follows from (c) and Lemma 19.3. Finally, to prove statement (e) we first note that if such an x exists then $k(X)^G = k$. On the other hand, if $k(X)^G = k$, then $d = \dim X$ (by (c)); we now apply statement (d). QED

Definition. Let B be a Borel subgroup of G. Let X be an irreducible variety on which G acts morphically. The *complexity* of X, denoted by $c(X)$, is the transcendence degree of $k(X)^B$ over k.

Lemma 19.7. *Let X be an irreducible variety on which G acts morphically. Let $B = TU$ be a Borel subgroup of G.*

(a) *$c(X) = \dim X - \max\{\dim B{\cdot}x : x \in X\}$.*
(b) *If $c(X) = 0$, then the action of G on k[X] is multiplicity-free. If the action of G on k[X] is multiplicity-free and if X is quasi-affine, then $c(X) = 0$.*
(c) *If $c(X) \geq 1$ and if X is quasi-affine, then there is a character $\omega \in X(B)$ such that for each positive integer n, $\dim \{f \in k[X] : b{\cdot}f = (n\omega)(b)f \text{ for all } b \in B\} \geq n + 1$.*
Proof. Statement (a) follows at once from Corollary 19.6(c) and (d). Now, let us prove statement (b). If $c(X) = 0$, then a Borel subgroup of G has an open orbit in X by statement (a) and we have already proved that such an action is multiplicity-free (in the paragraph following the statement of Theorem 11.1). Conversely, suppose that the action is multiplicity-free and let $f \in k(X)^B$. According to Lemma 19.1, $f = a/b$ where both a and b are in $k[X]^U$ and have the same T-weight. Since the action of G is multiplicity-free, a differs from b by an element in k and $f \in k$. Thus, $k(X)^B = k$ and $c(X) = 0$.

If $c(X) \geq 1$, then there is an element $f \in k(X)^B$ which is transcendental over k. According to Lemma 19.1, $f = a/b$ where both a and b are in $k[X]^U$ and have the same T-weight, say ω. For $i = 0, \ldots, n$, each of the elements $a^i b^{n-i}$ in $k[X]^U$ has T-weight $n\omega$. We show now that these elements are linearly independent over k. Indeed, suppose to the contrary that there are elements $c_i \in k$, not all 0, such that $\Sigma_i c_i a^i b^{n-i} = 0$. Let x and y be indeterminates over k. The polynomial $\Sigma_i c_i x^i y^{n-i}$ is homogeneous of degree n and, so, factors into the product of n terms, each of which has the form $\alpha x + \beta y$ for suitable $\alpha, \beta \in k$. Therefore, replacing x by a and y by b, we see that one of the terms $\alpha a + \beta b$ must be 0. But then $a/b \in k$ which is a contradiction. QED

Let $X(B)$ be the character group of B. Any **Z**-submodule of $X(B)$ is free and finitely generated; the number of elements in a basis is called the *rank* of the submodule. This observation allows us to make the following definition.

Definition. Let B be a Borel subgroup of G. Let X be an irreducible variety on which G acts morphically. Let $\Gamma(X) = \{\chi \in X(B) : \text{there exists an } f \in k(X) \text{ with } b{\cdot}f = \chi(b)f \text{ for all } b \in B\}$. The *rank* of X, denoted by $r(X)$, is the rank of $\Gamma(X)$.

Lemma 19.8 [78; Corollary 2, p.253]. *Let X be an irreducible affine or quasi-affine variety on which G acts morphically. Let B = TU be a Borel subgroup of G. Then, $\dim k(X)^U = c(X) + r(X)$.*
Proof. Any quasi-affine variety on which G acts can be embedded as an open set in an affine variety on which G acts [95; Lemma 2, p.217]. Thus, we shall assume that X is affine. Let $\chi_1,\ldots,\chi_p$ be a basis of $\Gamma(X)$ over **Z** and let $f_1,\ldots,f_p$ be functions in $k(X)$ so that $b{\cdot}f_i = \chi_i(b)f_i$ for all $b \in B$ and $i = 1,\ldots,p$. We shall show that (i) $k(X)^U = k(X)^B(f_1,\ldots,f_p)$ and (ii) $f_1,\ldots,f_p$ are algebraically independent over $k(X)^B$. The lemma then follows by calculating the transcendence degrees on both sides of the preceding equation. To prove (i), let $f \in k(X)^U$. Then, $f = a/b$ where $a,b \in k[X]^U$ according to Corollary 19.2. Both a and b are sums of T-weight vectors since T normalizes U. Thus, in proving (i), we may assume that $f \in k[X]^U$ and $t{\cdot}f = \chi(t)f$ for some character $\chi \in X(T)$. Since $\chi \in \Gamma(X)$, we have $\chi = e(1)\chi_1 + \ldots + e(p)\chi_p$ for suitable integers $e(i)$. Then, $f_1^{e(1)} \ldots f_p^{e(p)}/f \in k(X)^B$. This proves statement (i). To prove (ii), suppose that there is a relation $\Sigma c_{(e)} f_1^{e(1)} \ldots f_p^{e(p)} = 0$, where each $c_{(e)} \in k(X)^B$. Each monomial $f_1^{e(1)} \ldots f_p^{e(p)}$

is a T-weight vector. Since $\chi_1, \ldots, \chi_p$ form a basis of $\Gamma(X)$, these monomials correspond to distinct weight vectors. But vectors corresponding to distinct weights are linearly independent. (We note that $t \cdot c_{(e)} = c_{(e)}$ for all $t \in T$ and each $c_{(e)}$ since $c_{(e)} \in k(X)^B$.) QED

Appendix. Our purpose in this appendix is to prove some facts about quotient spaces, including Theorem 19.5 and a theorem of Pommerening.

• In this appendix, we shall always assume (unless otherwise noted) that G is connected and that X is an irreducible variety on which G acts morphically.

We need a number of results which, for convenience, we place here. The following may all be found in [10].

(1) [Proposition 2.4, p.4] *Let G be a group of automorphisms of a field F. Then F is a separable extension of F^G, the field consisting of elements fixed by G.*

(2) [Proposition 6.6, p.97] *Suppose that $\pi: V \to W$ is a separable orbit map, and assume that W is normal and that the irreducible components of V are open. Then (W, π) is a quotient of V by G.*

(3) [Corollary, p.44] *Let $\alpha: V \to W$ be a dominant morphism of varieties, where W is normal. Assume that the dimensions of the irreducible components of the fibers of α are constant. Then α is an open map.*

(4) [Proposition, p.43] *Let $\alpha: V \to W$ be a dominant morphism of irreducible varieties, and suppose that $f \in k[V]$ is constant along the fibers of α. Then f is purely inseparable over $k(W)$.*

(5) [Lemma, p.43] *Let x be a normal point on an irreducible variety W and suppose that $h \in k(W)$ is not defined at x. Then there is a $y \in W$ at which $1/h$ is defined and vanishes.*

(6) [Proposition 6.15, p.102] *Let G be a finite group and let V be an affine variety on which G acts morphically. Then V/G exists and is affine.*

Of great importance is the following theorem of Chevalley.

(7) [17; Théorème 3, p.115]. *Let $\varphi: V \to W$ be a dominant morphism of irreducible varieties and let φ^o be its comorphism. Let R be the largest subfield of $k(V)$ which contains $\varphi^o(k(W))$ and is algebraic over $\varphi^o(k(W))$. Then R is a finite extension of $\varphi^o(k(W))$. Let n be the degree of the largest separable extension of $\varphi^o(k(W))$ contained in R. Then, there is a non-empty open subset W' of W such that for each $y \in W'$, $\varphi^{-1}(y)$ has exactly n irreducible components.*

To prove the first lemma below, it is convenient to introduce some terminology. Let k,

K, and L be subfields of a field F. Suppose that both K and L contain k. We say that K and L are *linearly disjoint over* k if the mapping from $K \otimes_k L$ to F given by $\Sigma_i a_i \otimes b_i \to \Sigma_i a_i b_i$ is injective. In particular, if K and L are linearly disjoint over k, then any set of elements in K (resp. L) which is linearly independent over k is also linearly independent over L (resp. K).

Lemma 1. *Let K be a finitely generated extension field of k. Let E be a subfield of L which contains k. Then, E is a finitely generated extension field of k.*
Proof. Let $y_1,\ldots,y_r,\ldots,y_s$ be a transcendence base for K over k such that $y_1,\ldots,y_r$ is a transcendence base for E over k. We shall show that E is a finite (algebraic) extension field of $k(y_1,\ldots,y_r)$. The field K is a finite (algebraic) extension field of $k(y_1,\ldots,y_s)$ since K is a finitely generated extension field of k and is algebraic over $k(y_1,\ldots,y_s)$.

Next, we show that the fields E and $k(y_1,\ldots,y_s)$ are linearly disjoint over $k(y_1,\ldots,y_r)$. If this were not so, then the elements $y_{r+1},\ldots,y_s$ would be algebraically dependent over E. The transcendence degree of $E(y_1,\ldots,y_s)$ over $k(y_1,\ldots,y_r)$ is the sum of the transcendence degrees of $E(y_1,\ldots,y_s)$ over E and of E over $k(y_1,\ldots,y_r)$; the second term is 0 while the first, by assumption, is $< s - r$. On the other hand, the transcendence degree of $E(y_1,\ldots,y_s)$ over $k(y_1,\ldots,y_r)$ is the sum of the transcendence degrees of $E(y_1,\ldots,y_s)$ over $k(y_1,\ldots,y_s)$ and of $k(y_1,\ldots,y_s)$ over $k(y_1,\ldots,y_r)$; the second term is $s - r$ while the first is 0. This contradiction shows that E and $k(y_1,\ldots,y_s)$ are linearly disjoint over $k(y_1,\ldots,y_r)$. We may conclude that any set of elements in E which is linearly independent over $k(y_1,\ldots,y_r)$ is also linearly independent over $k(y_1,\ldots,y_s)$.

We now show that E is finite over $k(y_1,\ldots,y_r)$. If this is not so, then E is an infinite dimensional vector space over $k(y_1,\ldots,y_r)$. By the linear disjointness proved in the preceding paragraph, there are infinitely many elements in K which are linearly independent over $k(y_1,\ldots,y_s)$. But this contradicts the fact that K is a finite (algebraic) extension of $k(y_1,\ldots,y_s)$. QED

Note. Let K be a finitely generated extension field of k and let G be a group of automorphisms of K over k. Then, the fixed field $K^G = \{f \in K : g{\cdot}f = f \text{ for all } g \in G\}$ is a finitely generated extension field of k according to Lemma 1. In this sense, the "finite-generation of invariants" question is trivial in the context of fields. It is interesting to note that Cayley and Sylvester always insisted on the *polynomial* character of invariant theory [110; p.234].

Lemma 2. *There is a dense, open, G-invariant subset X' in X so that $\Gamma = \{(x,y) \in X' \times X' :$ there is a $g \in G$ with $y = g{\cdot}x\}$ is closed in $X' \times X'$.*
Proof. Applying Lemma 19.3, we may assume that all of the G-orbits on X have the same dimension and, thus, are closed. Now, let ψ: $G \times X \to X \times X$ be given by $(g,x) \to (x,g{\cdot}x)$. The group G acts on $G \times X$ by $g{\cdot}(g_1,x) = (gg_1,x)$ and on $X \times X$ by $g{\cdot}(x,y) = (x,g{\cdot}y)$. The mapping ψ is G-equivariant with respect to these actions. Let the closure of the image of ψ be Z; there is a G-invariant set V in the image which is open in Z. Let p: $X \times X \to X$ be given by $p(x,y) = x$. Consider the restriction of p to V; for $v \in V$ with $p(v) = x$, we have $p^{-1}(p(v)) = p^{-1}(x) = (x, G{\cdot}x)$ an irreducible set of dimension m, say. In the terminology of preliminary (7), we have $n = 1$. We apply (7) again to the mapping p: $Z \to X$ recalling that $k(Z) = k(V)$. It follows that there is a dense, open G-

invariant subset X' of X such that $p^{-1}(x)$ has exactly one irreducible component of dimension m which must be $(x,G{\cdot}x)$ (see Exercise 3). To complete the proof, let (x,y) be in the closure of Γ. Then $(x,y) \in Z$ and $p(x,y) = x \in X'$. This implies that $(x,y) \in p^{-1}(x) = (x,G{\cdot}x)$. QED

Lemma 3. (a) *Suppose that the quotient X/G exists. Then there is a dense, open, G-invariant subset $X' \subset X$ and functions $f_i \in k[X']^G$, for $i = 1, \dots ,r$, which separate the G-orbits on X'.*
(b) *Suppose that there is a dense, open, G-invariant subset $X' \subset X$ and functions $f_i \in k[X']^G$, for $i = 1, \dots ,r$, which separate G-orbits on X'. Then, there is a dense, open, G-invariant subset $X'' \subset X$ such that the quotient X''/G exists.*
(c) *Let Y be any normal, irreducible variety on which G acts regularly. Let $\varphi: X \to Y$ be a G-equivariant morphism which is injective and dominant. If the quotient X/G exists, then there is a dense, open, G-invariant subset $Y'' \subset Y$ such that the quotient Y''/G exists.*
Proof. To prove (a), let $\pi: X \to X/G$ be the quotient mapping. Let Y be a dense, open, affine subset in X/G and let $X' = \pi^{-1}(Y)$. Let $k[Y] = k[f_1, \dots ,f_r]$. Then, the functions f_i and the subset X' have the desired properties.

Now, we prove statement (b). We may assume that X is normal. According to Lemma 1, the field $k(X)^G$ is finitely generated. By adding the generators of this field to the f_i and then removing their poles from X' to obtain X'', we may assume that the f_i satisfy $k(X')^G = k(X'')^G = k(f_1, \dots ,f_r)$. Let Y be the normalization in $k(X)^G$ of the affine variety corresponding to $k[f_1, \dots f_r]$. Let $\pi: X'' \to Y$ be the natural mapping. Then, π is a separable (by preliminary (1)) orbit map and, so, is a quotient mapping by preliminary (2).

Finally, we prove statement (c). The image of φ in Y is open by (3). Hence, we may assume that φ is surjective. Let $f \in k[X]$. Then, by preliminary (4), there is a positive integer m (a power of char k) so that $f_1 = f^m \in k(Y)$. We show now that f_1 is in $k[Y]$. If not, then f_1 has a pole W in Y since Y is normal. We now apply (5); let $w \in W$ be a point where $1/f_1$ is defined and $(1/f_1)(w) = 0$. Let $w = \varphi(x)$. Then, $0 = (1/f_1)(w) = (1/f_1)(\varphi(x)) = (1/f_1)(x)$. But, this is impossible since $1 = f_1(x)\cdot(1/f_1(x))$. By statement (a), we may now find a dense, open, G-invariant subset $X' \subset X$ and functions $f_i \in k[X']^G$ for $i = 1, \dots ,r$ which separate the G-orbits on X'. Then, $Y' = \varphi(X')$ is open in Y and the G-orbits on Y' are separated by powers of the f_i. We apply statement (b) to complete the proof. QED

Lemma 4. *Let X be non-singular and suppose that all of the G-orbits on X have the same dimension. Let $x_o \in X$. Let X_o be an affine open subset of X such that $x_o \in X_o$ and $(G{\cdot}x_o \cap X_o)$ is a proper subset of X_o. We consider X_o to be a subset of some r-dimensional affine space A^r. Then, there is a linear subspace L of A^r such that (i) $L \cap (G{\cdot}x_o \cap X_o)$ is finite and non-empty; (ii) $U_l = L \cap X_o$ is irreducible, normal, and affine; (iii) $\dim U_l = \dim X - \dim G{\cdot}x_o$. (Changing notation, we shall assume from now on that $x_o \in L \cap (G{\cdot}x_o \cap X_o)$.)*
Proof. This is really a fact about generic hyperplane sections and we give the proof assuming the known results [66; Section 6, p.211]. The subspace $Y = G{\cdot}x_o \cap X_o$ is closed in X_o. Suppose that X_o and Y are defined over some subfield K of k and let x and

y be generic points over K in k of X_0 and Y, respectively. Let $t_1, \ldots, t_{r+1}$ be $r+1$ algebraically independent elements in k over $K(x,y)$. Let H_t be the "generic hyperplane" defined by $t_1X_1 + \ldots + t_rX_r = t_{r+1}$. It is known that (i) if dim $Y \geq 2$, then both $H_t \cap X_0$ and $H_t \cap Y$ are non-singular, irreducible and non-empty; (ii) if dim $Y \geq 1$, then both $H_t \cap X_0$ and $H_t \cap Y$ are non-empty; (iii) $\dim(H_t \cap X_o) = \dim X_0 - 1$ and $\dim(H_t \cap Y) = \dim Y - 1$. By repeating this process, we obtain the desired subspace L. QED

Proof of Theorem 19.5. We now prove Rosenlicht's theorem in six steps. The outline is this. We shall find a variety D_1'' on which G acts regularly and a G-equivariant mapping p_1 from D_1'' to an open subset X'' of X which is m:1 for some positive integer m. Furthermore, there is a finite group H which acts on D_1'' and permutes the m points above any $x \in X''$. We show that the quotient D_1''/H exists and then that there is an open subset U_2 of D_1''/H so that the quotient U_2/G exists. The proof is then finished by applying Lemma 3(c) to the mapping $U_2 \to X$.

To start, we apply Lemma 19.3 and Lemma 2 to see that we may replace X by a non-empty open subvariety and always assume that (i) X is a non-singular, (ii) all of the G-orbits on X have the same dimension and, thus, are closed, and (iii) $\Gamma = \{(x,y) \in X \times X : \text{there is a } g \in G \text{ with } y = g{\cdot}x\}$ is closed in X. Let U_1 and x_0 be as in Lemma 4. We should note that if $G{\cdot}x_0 \cap X_0 = X_0$, then $G{\cdot}x_0$ is open and closed in X so $G{\cdot}x_0 = X$ and the theorem is immediate.

(I) *Let* $\Gamma_1 = (U_1 \times X) \cap \Gamma = \{(u_1, g{\cdot}u_1) : u_1 \in U_1, g \in G\}$. *Then,* Γ_1 *is a closed subset of* $U_1 \times X$ *since* Γ *is closed. The group* G *acts on* Γ_1 *by* $g{\cdot}(u_1,x) = (u_1, g{\cdot}x)$. *Let* $\pi\colon \Gamma_1 \to U_1$ *be given by* $\pi(u_1,x) = u_1$. *Then,* π *is a quotient mapping.*
Proof. First, the mapping π is surjective. Second, according to (3), it is open since its fibers are orbits all of the same dimension. Finally, we must show that if V is an open G-invariant subset of Γ_1, then π^o induces an isomorphism from $k[\pi(V)]$ to $k[V]^G$. Let id: $U_1 \to \Gamma_1$ be given by $u_1 \to (u_1, u_1)$. Let $F \in k[V]^G$ and put $f = F \circ id$. Then, $f \in k[\pi(V)]$ and $(\pi^o f)(u_1, g{\cdot}u_1) = f(u_1) = F(u_1,u_1) = F(u_1, g{\cdot}u_1)$. QED

(II) *Let* $p\colon \Gamma_1 \to X$ *be given by* $p(u_1, g{\cdot}u_1) = g{\cdot}u_1$. *Then,* p *is* G*-equivariant, dominant, and finite.*
Proof. By (I) and Lemma 4, we have $\dim \Gamma_1 = \dim U_1 + \dim (G{\cdot}x_o) = (\dim X - \dim G{\cdot}x_o) + \dim G{\cdot}x_o = \dim X$. Also, $p^{-1}(x_0) = \{(u_1,x_0) : u_1 \in U_1\}$. Such u_1 must be in $G{\cdot}x_0 \cap U_1$ which is finite by Lemma 4. QED

According to preliminary (7), there is an integer n and a dense, open, G-invariant subset $X' \subset X$ so that $p^{-1}(x)$ has exactly n elements for all $x \in X'$. We replace X by X'. Since the inverse image of X' is G-invariant, it has the form $(U_1' \times X) \cap \Gamma$. The set U_1' is open in U_1 since it is $\pi(p^{-1}(X'))$. We replace U_1 by U_1'. Thus, we may assume that p is an n:1 mapping. Next, consider the mapping ψ: $\Gamma_1^n = \Gamma_1 \times \ldots \times \Gamma_1$ (n-copies) $\to X^n = X \times \ldots \times X$ given by $((u_1, g_1{\cdot}u_1), \ldots, (u_n, g_n{\cdot}u_n)) \to (g_1{\cdot}u_1, \ldots, g_n{\cdot}u_n)$. Let $D_X = \{(x, \ldots, x) \in X^n\}$ and let D be its inverse image under ψ. There is a mapping p_{ij}: $D \to \Gamma_1 \times \Gamma_1$, $(z_1, \ldots, z_n) \to (z_i, z_j)$. The inverse image under this mapping of the diagonal is closed in D. We remove these G-invariant, closed subsets from D and rename the resulting set D. Thus, $((u_1, g_1{\cdot}u_1), \ldots, (u_n, g_n{\cdot}u_n)) \in D$ means that the $(u_i, g_i{\cdot}u_i)$, $i = 1$,

... $,n$, are precisely the n distinct points in Γ_1 lying over $x = g_1 \cdot u_1 = \ldots = g_n \cdot u_n$.

Let p_1: $D \to X$ be given by $p_1((u_i, g_i \cdot u_i)) = g_1 \cdot u_1$. The group G acts on D by $g \cdot ((u_i, g_i \cdot u_i)) = ((u_i, gg_i \cdot u_i))$ and p_1 is a G-equivariant map. The symmetric group on n letters, S_n, also acts on D by permuting the co-ordinates $(u_i, g_i \cdot u_i)$. This action commutes with that of G. Furthermore (and this is very important), for each $x \in X$, the $n!$ points in $p_1^{-1}(x)$ are permuted by S_n.

Let $D = D_1 \cup \ldots \cup D_r$ be the decomposition of D into irreducible components. For each $i = 1, \ldots, r$, let $D_i' = D_i - \cup (D_i \cap D_j)$ where i and j are distinct; the D_i' are pairwise disjoint. Let $D' = D_1' \cup \ldots \cup D_r'$. Then, D' is open and dense in D. Its irreducible, connected components are the D_i' [10; Proposition 1.2, p.2]. Each D_i' is G-invariant. Furthermore, S_n maps D' to itself and thus permutes the D_i'. Since p_1 is a dominant and finite mapping, we may assume that its restriction to D_1' is also dominant and finite. Replacing D_1' by a smaller G-invariant subset D_1'', we may assume that D_1'' is normal and that p_1 is an m:1 mapping on D_1'' for some positive integer m. The set X'' $= p_1(D_1'')$ is dense, open, and G-invariant in X. Let H be the subgroup consisting of all $\sigma \in S_n$ such that $\sigma(D_1'') = D_1''$.

(III) *Let $x \in X''$ and let $p_1^{-1}(x) = \{d_1, \ldots d_m\} \subset D_1''$. For each i,j such that $1 \leq i < j \leq m$, there is a $\sigma \in H$ such that $\sigma(d_i) = d_j$.*

Proof. Let us take $i = 1$ and $j = 2$. We have noted above that there is a $\sigma \in S_n$ so that $\sigma(d_1) = d_2$. Suppose that $\sigma(D_1') = D_j'$ where $j > 1$. Then, $d_2 \in \sigma D_1' \cap D_1' = D_j' \cap D_1'$ which is impossible since these two sets are disjoint. QED

(IV) *The quotient D_1''/H exists.*

Proof. The mapping p_1: $D_1'' \to X''$ is an open m:1 mapping. Thus, D_1'' is the normalization of X'' in $k(D_1'')$ by Theorem 1.10. Let X' be the normalization of X'' in $k(D_1'')^H$. Applying preliminary (4), we see that the mapping from X' to X'' is injective since $k(D_1'')^H$ is purely inseparable over $k(X'')$. Then p_1: $D_1'' \to X'$ is a separable, orbit mapping for H whose image is normal. It follows from preliminary (2) that D_1''/H exists. QED

The group H acts on the affine variety U_1^n and the quotient U_1^n/H exists by preliminary (6). Consider π: $D_1''/H \to U_1^n/H$, $((u_i, g_i \cdot u_i)) \to (u_i)$. Obviously, π is constant on G-orbits.

(V) *The mapping π separates G-orbits.*

Proof. Let $u = ((u_i, g_i \cdot u_i))$ and $v = ((v_i, h_i \cdot v_i))$ be points in D_1''/H having the same image in U_1^n/H. The point u lies above $x = g_1 \cdot u_1 = \ldots = g_n \cdot u_n$ and v above $y = h_1 \cdot v_1 = \ldots$ $= h_n \cdot v_n$. Furthermore, the u_i's are just a permutation of the v_j's. Suppose that $v_2 = u_1$. Then, $p_1(h_2 g_1^{-1} \cdot ((u_1, g_1 \cdot u_1), \ldots)) = p_1((u_1, h_2 \cdot u_1), \ldots) = p_1((v_2, h_2 \cdot v_2), \ldots) = h_2 \cdot v_2 = y =$ $p_1(v)$. By (III), there is a $\sigma \in H$ so that $\sigma(h_2 g_1^{-1} \cdot u) = v$. Since the actions of G and S_n commute, we see that $h_2 g_1^{-1} \sigma u = v$. Thus, u and v are in the same G-orbit on D_1''/H. QED

(VI) Since U_1^n/H is affine, we may separate G-orbits on D_1''/H by finitely many G-invariant functions. Thus, on an open, dense G-invariant subset U_2 of D_1''/H, the quotient U_2/G exists (by Lemma 3(b)). But, the mapping $D_1''/H \to X''$ is injective and G-equivariant. Thus, on an open, dense G-invariant subset X' of X, the quotient X'/G exists by Lemma 3(c). QED

Note. In proving Rosenlicht's theorem, we have generally followed the argument in [99]. Other proofs may be found in [96] and [101].

Theorem 5 [85; Theorem, p. 12]. *Let X be an irreducible, affine variety such that $k[X]$ is a unique factorization domain and let G (not necessarily connected) act morphically on X. Then, there is a dense, open subset X' of X such that for each $x \in X'$, the stabilizer G_x is a Grosshans subgroup of G.*

Proof. We may assume that G is connected (Corollary 4.4). According to Lemma 1, there are finitely many functions $f_1, \dots, f_r \in k(X)^G$ such that $k(X)^G = k(f_1, \dots, f_r)$. Suppose that f_1 has poles at the subvarieties $P_1, \dots, P_s$. Since f_1 is G-invariant, so are each of the P_i. Hence, there are irreducible G-weight functions $p_i \in k[X]$ such that $P_i = \{x \in X : p_i(x) = 0\}$. Let $p_1, \dots, p_q$ be all the irreducible functions obtained in this way from $f_1, \dots, f_r$ and let $f = p_1 \dots p_q$. For each $i = 1, \dots, r$, there is a non-negative integer $e(i)$ such that $f^{e(i)}f_i \in k[X]$. Furthermore, the open subvariety $X_f = \{x \in X : f(x) \neq 0\}$ is G-invariant, $k(X_f)^G = k(X)^G$, and each $f_i \in k[X_f]^G$. Let Y be the affine variety such that $k[Y] = k[f_1, \dots, f_r]$ and let $\pi: X_f \to Y$ be the natural mapping.

Let $m = \max\{\dim G{\cdot}x : x \in X\}$. According to Lemma 19.3, the set $Z = \{x \in X_f : \dim G{\cdot}x = m\}$ is G-invariant and open in X_f. Since $k[X]$ is a unique factorization domain, the (G-invariant) subvariety of $X_f - Z$ consisting of those components of codimension 1 is the zeros of some function $h \in k[X]$. Let W be the union of the components of $X_f - Z$ having codimension > 1. The subvariety $X_{fh} = Z \cup W$ is G-stable; from now on, we restrict π to X_{fh}. We note that π is dominant. The fibers of π are G-invariant and for all $x \in Z$, we have $\pi^{-1}(\pi(x)) = (\pi^{-1}(\pi(x)) \cap Z) \cup (\pi^{-1}(\pi(x)) \cap W)$. We now show that there is a non-empty, open subset $Y' \subset Y$ such that for each $y \in Y'$, $\dim \pi^{-1}(y) = m$ and $\dim(\pi^{-1}(y) \cap W) \leq m - 2$. First, there is a dense, open subset $Y' \subset Y$ such that $\pi^{-1}(y)$ consists only of components having dimension m ([10; Theorem (2), p.21] and Corollary 19.6(c)). If the closure of $\pi(W)$ is a proper subset of Y we replace Y' by $Y' \cap (Y - \mathrm{cl}(\pi(W)))$; thus, $\pi^{-1}(y) \cap W$ is empty. Otherwise, the closure of $\pi(W) = Y$. Restricting π to W for a moment, we see that there is an open, dense subset Y'' such that for each $y \in Y''$, $\dim \pi^{-1}(y) = \dim W - \dim Y \leq \dim X - 2 - (\dim X - m) = m - 2$. We replace Y' by $Y' \cap Y''$.

Now, $Z'' = Z \cap \pi^{-1}Y'$ is G-invariant, open, and dense in X. Let $x \in Z''$. Then $X_1 = \mathrm{cl}(G{\cdot}x) \cap X_{fh}$ is an irreducible component of $\pi^{-1}(\pi(x))$ since all G-orbits on X_{fh} have dimension m. Furthermore, the set $X_1 - G{\cdot}x$ does not contain any point in Z since the orbit of such a point has dimension m. Hence, $X_1 - G{\cdot}x \subset \pi^{-1}(\pi(x)) \cap W$ and $\dim(X_1 - G{\cdot}x) \leq \dim(\pi^{-1}(\pi(x)) \cap W) \leq m - 2 \leq \dim X_1 - 2$. Therefore G_x is a Grosshans subgroup of G. QED

Exercises.

1. Let GL_n act on M_n, the algebra consisting of all $n \times n$ matrices over k by $g{\cdot}x = gxg^{-1}$.

(a) For $x \in M_n$, let $c_1(x), \dots, c_n(x)$ be the coefficients of the characteristic polynomial

of x. Show that the c_i are G-invariant polynomials on M_n.
(b) Show that the algebra of GL_n-invariant polynomials on M_n is the polynomial algebra generated by $c_1, \ldots, c_n$. (Hint: Let D be the set of diagonal matrices in M_n. Show that the restriction mapping from M_n to D gives an isomorphism between GL_n-invariant polynomials and symmetric functions in n variables.)
(c) In the context of Theorem 19.5, show that the set X' consists of all those matrices x having distinct eigenvalues.

2. Let $B = TU$ be a Borel subgroup of G. Let G act morphically on an irreducible variety X in such a way that $c(X) = 0$. Let X' be a non-empty open subset of X such that X'/U exists. Show that we may assume that X' is T-invariant and that T has an open orbit on X'/U.

3. Prove that the set X' defined in the proof of Lemma 2 is G-invariant. (Hint: G acts on the image of ψ and, then, Z by $g^*(x,y) = (g{\cdot}x, g{\cdot}y)$. Then, $p^{-1}(g{\cdot}x) = g^*{\cdot}p^{-1}(x)$.)

4. Let $B = TU$ be a Borel subgroup of G. Let G act morphically on an irreducible variety X in such a way that $c(X) = 0$. Let ω be a character of B and suppose that f_1, f_2 are non-zero functions in $k(X)$ such that $b{\cdot}f_i = \omega(b)f_i$ for all $b \in B$. Show that there is an element $c \in k$ with $f_1 = cf_2$.

§20. Unique Factorization Domains

In this section, we shall always assume that $k[X]$ is a unique factorization domain. For such actions, there are relatively easy proofs of the general theorems to be presented in §21; furthermore, these arguments give somewhat more information than we obtain in the general case. In what follows, we shall always assume that $B = TU$ is a Borel subgroup of G.

A. $c(X) = 0$

Lemma 20.1. *Let A be a unique factorization domain whose group of units is k^* and on which G acts rationally. If G is connected and has no non-trivial character, then A^G is a unique factorization domain.*

Proof. Let $a \in A^G$ be a non-zero, non-unit element and let $a = p_1 \ldots p_r$ be its prime factorization. Then, for each $g \in G$, we have $a = g{\cdot}a = (g{\cdot}p_1) \ldots (g{\cdot}p_r)$ and the $g{\cdot}p_i$ are the p_j up to constants and rearrangement. Since G is connected, $g{\cdot}p_i = cp_i$, where $c \in k^*$. Since G has no non-trivial character, $g{\cdot}p_i = p_i$. QED

In case the complexity of the action of G on X is 0, we are able to describe the algebra of invariants and also its relationship to the orbit structure of G on X. The next definition is the key to understanding this.

Definition. Let G act morphically on the variety X. An orbit $G{\cdot}x$ is said to be G-*separated* if $G{\cdot}x = \{y \in X : f(y) = f(x) \text{ for all } f \in k[X]^G\}$.

It is known that if X is quasi-affine and G is unipotent, then there is a non-empty subset X'' of X consisting of U-separated orbits (Exercise 1).

Theorem 20.2. *Let X be an irreducible, quasi-affine variety on which B acts morphically such that $c(X) = 0$. Suppose that $k[X]$ is a unique factorization domain with unit group k^*.*
(a) *$k[X]^U$ is a finitely generated polynomial algebra, say, $k[X]^U = k[p_1, \dots, p_r]$ where p_i is a B-weight function.*
(b) *If $x \in X$ is such that each $p_i(x) \neq 0$, then the orbit $U{\cdot}x$ is U-separated.*
Proof. The action of B on $k[X]$ is multiplicity-free (Lemma 19.7(b)). Let $\{\chi_1, \dots, \chi_p\}$ be a **Z**-basis for $\Gamma(X)$. For each $i = 1,2, \dots ,p$, let f_i be an element in $k(X)$ so that $b{\cdot}f_i = \chi_i(b)f_i$ for all $b \in B$. For any $\chi \in \Gamma(X)$, say $\chi = e(1)\chi_1 + \dots + e(p)\chi_p$, the element $f = f_1^{e(1)} \dots f_p^{e(p)}$ is the unique (up to scalar) function in $k(X)$ with $b{\cdot}f = \chi(b)f$ (Exercise 4, §19). As in Corollary 19.2, we write $f_i = a_i/b_i$ where a_i and b_i are T-weight vectors in $k[X]^U$. Let $p_1, \dots, p_r$ be all the distinct primes appearing in the prime factorizations of $a_1, \dots, a_p, b_1, \dots, b_p$. We show now that $k[X]^U = k[p_1, \dots, p_r]$. Indeed, let $f \in k[X]^U$ have T-weight χ. As noted above, $f = cf_1^{e(1)} \dots f_p^{e(p)}$ for some $c \in k$. Writing f_i as a_i/b_i and collecting terms on both sides of the previous equation, we see that there are monomials m,m' in the a_i and b_i so that $m'f = m$. Thus, the primes appearing in the prime factorization of f are among the $p_1, \dots, p_r$.

Next, we shall show that the p_i's are algebraically independent. Let p_i have T-weight μ_i. Then $p_1^{e(1)} \dots p_r^{e(r)}$ is the only monomial in the p's with T-weight $e(1)\mu_1 + \dots + e(r)\mu_r$. Indeed, suppose that $p_1^{d(1)} \dots p_r^{d(r)}$ has this T-weight. Since $c(X) = 0$, there is a $c \in k^*$ with $p_1^{e(1)} \dots p_r^{e(r)} = cp_1^{d(1)} \dots p_r^{d(r)}$ and we may conclude (using unique factorization) that $e(1) = d(1), \dots, e(r) = d(r)$. Therefore, any relation of the form $\Sigma c_{(e)}p_1^{e(1)} \dots p_r^{e(r)} = 0$ must have each $c_{(e)} = 0$ since the various monomials have different T-weights. This completes the proof of statement (a).

To prove statement (b), let Y be the affine variety such that $k[Y] = k[X]^U = k[p_1, \dots, p_r]$. Then, $k[Y]$ is a unique factorization domain according to Lemma 20.1. Furthermore, the injection $k[Y] \subset k[X]$ gives a T-equivariant mapping π: $X \to Y$. Thus, for each $y \in Y$ and $t \in T$, we have $\pi^{-1}(t{\cdot}y) = t{\cdot}\pi^{-1}(y)$. We show first that there is a point $y_0 \in Y$ such that the orbit $T{\cdot}y_0$ is open in Y. For otherwise, there is a non-constant T-invariant rational function $a/b \in k(Y)$ by Lemma 19.7 applied to T and Y. We may assume that both a and b are in $k[Y]$ and have the same T-weight χ (Lemma 19.1). But $\dim k[X]^U_\chi \leq 1$ and, so, $a/b \in k$.

The subvariety $Y - T{\cdot}y_0$ is the union of irreducible subvarieties $Y_1, \dots, Y_p$, each having codimension 1 (according to the proof of Corollary 2.5). Since $k[Y]$ is a unique factorization domain, we may find irreducible T-weight functions $a_1, \dots, a_p$ in $k[Y]$ such that $Y_i = \{y \in Y : a_i(y) = 0\}$ for $i = 1, \dots ,p$. But $k[Y] = k[p_1, \dots ,p_r]$ so each a_i is a monomial in the p_j's since $c(X) = 0$. It follows that we may take each a_i to be some p_j. (Also, each p_j must appear among the a_i's - Exercise 2.) Let X'' be a non-empty open subset of X consisting of U-separated orbits. The image $\pi(X'')$ intersects the orbit $T{\cdot}y_0$. For any y in the intersection, $\pi^{-1}(y)$ is an orbit $U{\cdot}x$. But, $\pi^{-1}(t{\cdot}y) = t{\cdot}\pi^{-1}(y) = t{\cdot}Ux = U{\cdot}tx$ and, so, $\pi^{-1}(T{\cdot}y_0)$ consists entirely of U-separated orbits. QED

Example. We shall show that the "co-adjoint representation" satisfies the conditions of Theorem 20.2. In this example, let char $k = 0$ and let G be semi-simple. We shall use the notation introduced at the beginning of §3 and, in particular, let $B = TU$ be a Borel subgroup of G. We shall take as positive roots those in $\Phi(T,U)$. Let L(G) be the Lie algebra of G; let $L(U)$ be the Lie algebra of U, on which the group B acts via the adjoint representation. We shall denote by e_α the element in $L(U)$ corresponding to a root $\alpha \in \Phi(T,U)$. Let V be the dual space of $L(U)$ together with the natural action by B; this is the *co-adjoint representation*. If we take all the e_α as a basis for $L(U)$, the elements in the dual basis will be denoted by e_α'; we should note that e_α' has T-weight $-\alpha$. The purpose of this example is to prove that $c(V) = 0$.

We give the argument in a series of steps. The second is the block on which the argument proceeds. It requires several facts about highest roots which we state here for convenience [12; Chapitre VI]. In doing so, for roots α and γ, we denote the α-sequence through γ by $\gamma - q\alpha, \dots, \gamma + p\alpha$ and let $p - q = -n(\gamma,\alpha) = -2(\gamma,\alpha)/(\alpha,\alpha)$.

(1). (a) *Let α and γ be positive roots with $\alpha \neq \gamma$ and suppose that $[e_\alpha, e_\gamma] = 0$. The following statements are equivalent: (i) $\gamma - \alpha$ is not a positive root; (ii) both e_α and $e_{-\alpha}$ belong to the commutator of e_γ; (iii) $n(\gamma,\alpha) = 0$.*
(b) [12; Proposition 25, p.165]. *Let δ be a highest root of $L(U)$. For any positive root α other than δ, $n(\alpha,\delta) = 0$ or 1.*

(2). *There are T-weight vectors $e_{\beta(1)}', \dots, e_{\beta(r)}'$ in V so that if W is the subspace of V spanned by $e_{\beta(1)}', \dots, e_{\beta(r)}'$ and $w = c_1 e_{\beta(1)}' + \dots + c_{\beta(r)} e_r'$ with $c_i \in k$ and $c_1 \dots c_r \neq 0$, then:*
(a) *$u \cdot w \in W$ if and only if $u \cdot w = w$;*
(b) *$u \cdot w = w$ if and only if $u \in U_{\beta(1)} \dots U_{\beta(r)}$,*
(c) *the roots $\beta(1), \dots, \beta(r)$ are linearly independent over $\mathbf{Z}$.*

Proof. Let $\beta(1)$ be a highest root in $L(G)$ and let $e_{\beta(1)}$ and $e_{\beta(1)}'$ be the corresponding elements in $L(U)$ and V. Let C_1 be the semi-simple part of the commutator in $L(G)$ of $e_{\beta(1)}$. The semi-simple rank of C_1 is less than the semi-simple rank of G. Let $\beta(2)$ be a highest root in C_1 and let $e_{\beta(2)}$ and $e_{\beta(2)}'$ be the corresponding elements in $L(U)$ and V. We continue on in this manner. Since the semi-simple ranks steadily decrease, this process must terminate. Now, the $\beta(i)$ are obviously linearly independent so it suffices to prove statements (a) and (b).

Let $u \in U$ and $w \in W$ have the form given in (a) and suppose that $u \cdot w \in W$. Let $u = \Pi u_\alpha(c_\alpha)$ where α ranges over certain positive roots α and we assume that none of the c_α is 0. Let i be the smallest positive integer so that there is a u_α appearing in u with $(\alpha,\beta(i)) \neq 0$. Among such α, choose the smallest relative to the usual ordering; we call this root α. We shall suppose that α is not one of the $\beta(i)$. Then, applying u to $e_{\beta(i)}'$, we see that a vector having T-weight $-\beta(i) + \alpha$ appears. We now show that this weight does not appear when u is applied to any of the $e_{\beta(j)}'$. Then, u does not send w to an element in W. In fact, suppose that $-\beta(i) + \alpha = -\beta(j) + \Sigma c_\beta \beta$ where the β are certain roots appearing in u and $j > i$. Then, $\alpha = \beta(i) - (\beta(j) - \Sigma c_\beta \beta)$. Since $\beta(j) - \Sigma c_\beta \beta$ is a positive root, any simple root, say α_r, appearing in it with non-zero coefficient must also appear in $\beta(j)$. Now, $(\beta(i), \beta(j)) = 0$ so each $(\beta(i), \alpha_r) = 0$ by statement (1b). This shows that $(\beta(i), \beta(j) - \Sigma c_\beta \beta) = 0$. Then, $(\beta(i), \alpha) = (\beta(i), \beta(i))$ and $n(\alpha, \beta(i)) = 2$ contradicting (1)b. QED

(3) Let $U_1 = \Pi_\alpha U_\alpha$ where $\alpha > 0$ and $\alpha \neq \beta(i)$ for all i. Let $U_2 = \Pi U_{\beta(i)}$. According to (2)b, U_2 is a subgroup of U. From the fact that U is directly spanned by all the root groups U_γ, we see that $U = U_1U_2$. For $i = 1,2$, let p_i: $U \to U_i$ be the natural projection. The group U acts on U_1 by $u \cdot u_1 = p_1(uu_1)$, i.e., for all $u,u' \in U$, we have $(uu')\cdot u_1 = u\cdot(u'\cdot u_1)$. Indeed, we need only show that $p_1(uu'u_1) = p_1(up_1(u'u_1))$. But, $u(u'u_1) = u(p_1(u'u_1)p_2(u'u_1)) = (up_1(u'u_1))p_2(u'u_1)) = p_1(up_1(u'u_1))p_2(up_1(u'u_1))p_2(u'u_1)$. The second and third terms are both in U_2 and the desired result follows.

(4) *Let* $W' = \{w \in W : w = c_1e_{\beta(1)}' + ... + c_re_{\beta(r)}'$ *with* $c_i \in k$ *and* $c_1...c_r \neq 0\}$. *Let* Φ: $U_1 \times W' \to V$ *be given by* $\Phi(u_1,w) = u_1w$. *Then*
(a) *Φ is injective;*
(b) *the image of Φ, denoted by V', is open in V;*
(c) *Φ is a U-equivariant isomorphism of $U_1 \times W'$ onto V' where $u\cdot(u_1,w) = (u\cdot u_1,w)$ is the action given in (2) and where U acts naturally on V;*
(d) $k[U_1 \times W']^U = k[W'] = k[V']^U$;
(e) [26] $c(V) = 0$ *and* $k[V]^U$ *is a finitely generated polynomial algebra.*
Proof. If $\Phi(u_1,w_1) = \Phi(u_2,w_2)$, then $u_1\cdot w_1 = u_2\cdot w_2$ and $u_2^{-1}u_1w_1 = w_2$. Then, (a) follows from (2)b and direct spanning. Applying statement (a), we have dim $\Phi(U_1 \times W') =$ dim $U_1 +$ dim $W' =$ dim $U_1 +$ dim $U_2 =$ dim $U =$ dim V. That the image of Φ is open follows at once from Lemma 5.1. Then, Φ is an isomorphism according to Theorem 1.10. It is U-equivariant since $\Phi(u\cdot(u_1,w)) = \Phi(u\cdot u_1,w) = (u\cdot u_1)\cdot w = p_1(uu_1)\cdot w$. Applying statement (2)b, we see that this last term is $p_1(uu_1)p_2(uu_1)\cdot w = (uu_1)\cdot w = u\cdot(u_1\cdot w) = u\cdot\Phi(u_1,w)$.

Now we prove statement (d). Let $f \in k[W']$; then $F \in k[U_1 \times W']$ defined by $F(u_1,w') = f(w')$ is U-invariant. Conversely, if $F \in k[U_1 \times W']$ is U-invariant, we define $f(w) = F(e,w)$. Then, $F(u_1,w') = F(u_1^{-1}\cdot(u_1,w')) = F(e,w') = f(w')$. The group T acts on V' and W' in the natural way and on $U_1 \times W'$ by $t\cdot(u_1, w) = (tu_1t^{-1}, w)$; the isomorphisms just given are all T-equivariant. Since the $\beta(i)$ are linearly independent over **Z**, T has an open orbit in W' so $k(V')^B = k(W')^T = k$. QED

B. $c(X) = 1$

Before getting to the main result, we need to define some terms.

Definition [10; p.29]. Let X be an irreducible variety. Then X is said to be *unirational over k* if there is an injective homomorphism β: $k(X) \to L$, where L is a finitely generated purely transcendental extension of k.

Lemma 20.3. *Let X be a unirational variety over k such that* dim $X = 1$. *Then there is a* $t \in k(X)$ *such that* $k(X) = k(t)$.
Proof. As noted in [10; AG 13.7], there is a dominant morphism f of a Zariski open subset V of affine space into X. We consider the image of lines L (i.e., $L \cap V$) in V. If dim $f(L) = 1$, then $f|L$ is dominant and the lemma follows from Lüroth's theorem. Otherwise, the image under f of each line L is a point. But this cannot be. Since f dominates X, there are points P,Q in V so that $f(P) \neq f(Q)$. Let L be the line connecting P and Q. Then $f|L$ is not constant. QED

Definitions. Let $B = TU$ be a Borel subgroup of G. Let X be an irreducible variety on which G acts morphically. Let $\Gamma^+(X) = \{\chi \in X(B) :$ there is a non-zero $f \in k[X]$ with $b{\cdot}f = \chi(b)f$ for all $b \in B\}$. If $\omega \in \Gamma^+(X)$, then $m_\omega = \dim \{f \in k[X] : b{\cdot}f = \omega(b)f$ for all $b \in B\}$. If ω and $\omega' \in \Gamma^+(X)$, then $\omega > \omega'$ if $\omega - \omega' \in \Gamma^+(X)$. We say that $\Gamma^+(X)$ satisfies the *finite descendent condition* if for any $\omega \in \Gamma^+(X)$, there are only finitely many $\omega' \in \Gamma^+(X)$ with $\omega > \omega'$.

Theorem 20.4 [79; Theorem 2, p.154]. *Let X be a unirational, quasi-affine irreducible variety on which G acts morphically in such a way that $c(X) = 1$ and $k[X]^B = k$. Suppose that $k[X]$ is a unique factorization domain with unit group k^* and that $\Gamma^+(X)$ satisfies the finite descendent condition. Then, there is a unique $\omega \in \Gamma^+(X)$ such that*
(a) *$m_\omega = 2$;*
(b) *if $\mu \in \Gamma^+(X)$ with $\mu = e\omega + \delta$ where $\delta - \omega \notin \Gamma^+(X)$, then $m_\mu = e + 1$.*
Proof. The algebra $k[X]^U$ is a unique factorization domain by Lemma 20.1. The field $k(X)^B$ has transcendence degree 1 over k so, by Lemma 20.3, there is a $t^* \in k(X)^B$ such that $k(X)^B = k(t^*)$. (Applying Theorem 19.5, we see that there is a non-empty open subset X' of X so that the quotient X'/B exists and has function field $k(X)^B$.) We may write t^* as f/h where f and h are B-weight functions having the same B weight, say $\omega > 0$ (Lemma 19.1). Since $\Gamma^+(X)$ satisfies the finite descendent condition, we may assume that ω is minimal with respect to this property. Also, since $\dim k(t^*) = 1$, we see that f and h are algebraically independent over k. Now, let us show that ω has the desired properties.

First, suppose that $g \in k[X]$ is a non-zero B eigenvector having B weight ω. Then, $f/g \in k(X)^B = k(t^*)$ so

$$f/g = \frac{\sum_{i=0}^{s} a_i (f/h)^i}{\sum_{j=0}^{r} b_j (f/h)^j}.$$

Multiplying the fraction on the right-hand side by an appropriate power of h, we obtain homogeneous polynomials in both f and h, in the numerator and denominator; these expressions may be factored. Then, clearing the denominators and simplifying we obtain

$$g{\cdot}\prod_{i=1}^{m-1} (a_i f + b_i h) = \prod_{j=1}^{m} (c_j f + d_j h)$$

where all the factors $af + bh$, $cf + dh$ are different, i.e., do not differ by an element in k. (We note that each side has the same T-weight $m\omega$.) Suppose now that $m > 1$ and let p be an irreducible element in $k[X]$ which divides $a_1 f + b_1 h$. Then, p also divides $c_1 f + d_1 h$, say. This means that the element $t^{**} = [(a_1 f + b_1 h)/p]/[(c_1 f + d_1 h)/p]$ also generates $k(X)^B$; but the expressions in the numerator and denominator of t^{**} have lower B-weight than those appearing in t^*. This contradiction shows that $m = 1$ and that $m_\omega = 2$.

Now, let us prove statement (b). Given $\mu \in \Gamma^+(X)$, there is a maximal integer $e \geq 0$ so that $\delta = \mu - e\omega \in \Gamma^+(X)$ since $\Gamma^+(X)$ satisfies the finite descendent condition. It follows that $\delta - \omega \notin \Gamma^+(X)$. Let $g \in k[X]$ be any non-zero B-eigenvector having B-weight δ. The B-eigenvectors $f^i h^{e-i} g$, $i = 0, \dots, e$, are linearly independent and all have B-weight $\mu = e\omega + \delta$. We now show that any B-eigenvector in $k[X]$ having B-weight μ is a linear combination of these. Let $f' \in k[X]$ be a non-zero B-eigenvector having B-weight μ. Then, $f'/(f^e g)$ is in $k(X)^B = k(t^*)$. Reasoning as before, we have $f' \cdot \Pi_i(a_i f + b_i h) = \Pi_j(c_j f + d_j h) \cdot g$ where all the factors $af + bh$ and $cf + dh$ are distinct. Some irreducible factor of $a_1 f + b_1 h$ must divide one of the $c_j f + d_j h$, say, for otherwise g is divisible by $a_1 f + b_1 h$ a polynomial having weight ω, contradicting our choice of e. Then we obtain the same contradiction as before unless the product on the left-hand side is empty, i.e., unless f' is a linear combination of the $f^i h^{e-i} g$. QED

Theorem 20.5. *Let X be a unirational, quasi-affine irreducible variety on which G acts morphically in such a way that $c(X) = 1$ and $k[X]^B = k$. Suppose that $k[X]$ is a unique factorization domain with unit group k^* and that $\Gamma^+(X)$ satisfies the finite descendent condition. Then, $k[X]^U$ is finitely generated over k.*

Proof. Let $\{\chi_1, \dots, \chi_p\}$ be a $\mathbf{Z}$-basis for $\Gamma(X)$. For each $i = 1,2, \dots ,p$, let f_i be an element in $k(X)$ so that $b \cdot f_i = \chi_i(b) f_i$ for all $b \in B$. As in Lemma 19.1, we write $f_i = a_i/b_i$ where a_i and b_i are T-weight vectors in $k[X]^U$. Let $p_1, \dots , p_r$ be all the distinct primes appearing in the prime factorizations of $a_1, \dots , a_p, b_1, \dots , b_p$. Let ω and t^* be as in Theorem 20.4 and its proof. If μ is one of the finitely many weights in $\Gamma^+(X)$ such that $\mu < \omega$, then its multiplicity is 1; let f_μ be a corresponding non-zero B-weight vector. We adjoin the irreducible factors of all such f_μ to the set $\{p_1, \dots , p_r\}$ along with f and h and show now that $k[X]^U$ is generated by these elements. Let $f \in k[X]^U$ have T-weight χ, where $\chi = e(1)\chi_1 + \dots + e(p)\chi_p$. Then, B fixes $f/(f_1^{e(1)} \dots f_p^{e(p)})$ so this element is a rational function in t^*, say =

$$\frac{\sum_{i=0}^{s} a_i (f/h)^i}{\sum_{j=0}^{r} b_j (f/h)^j}.$$

Writing f_i as a_i/b_i and collecting terms on both sides of the previous equation, we see that there are monomials m,m' whose terms are the p_i or of the form $af + bh$ such that $m'f = m$. Any prime appearing in the prime factorization of f is either one of the p_i or divides an expression $af + bh$. In the latter case, the prime is either $af + bh$, itself, or must appear in one of the f_μ. QED

We now look at the important case where X is a homogeneous space. In these situations, the technical conditions stated as part of the assumptions in Theorems 20.2 and 20.4 will be satisfied. For example, when G is simply connected and semi-simple, the Bruhat decomposition shows that $k(G)$ is a purely transcendental extension of k. The next theorem responds to the issue of unique factorization. Verifications of the other assumptions are usually left as exercises.

Theorem 20.6 [86; Corollary, p.303]. *Let G be a simply connected, semi-simple algebraic group. Then $k[G]$ is a unique factorization domain.*

Theorem 20.7 [21]. *Let G be reductive and let $B = TU$ be a Borel subgroup of G. Let U_1 be a closed subgroup of U such that $\dim(U/U_1) = 1$. Then U_1 is a Grosshans subgroup of U.*

Proof. We may assume that G is semi-simple and simply connected. The subgroup U_1

is observable in G by Corollary 1.5. The group G acts by left translation on G/U_1 and the B^- orbit of eU_1 has codimension 1. Thus, the complexity of G/U_1 is 1 (Lemma 19.7) and we may apply Theorems 16.2 and 20.5 to see that $k[G/U_1]$ is finitely generated over k. QED

Theorem 20.8 [61; Korollar, p.36]. *Let G be reductive and let $B = TU$ be a Borel subgroup of G. Let H be a closed subgroup of U which is normalized by T. Let Ψ be the set of positive roots (relative to B) which are not in $\Phi(T,H)$ and let $<\Psi>_Q$ denote the vector space spanned by Ψ over Q. If $\dim<\Psi>_Q \geq \text{Card}\ \Psi - 1$, then H is a Grosshans subgroup of G.*

Proof. We may assume that G is simply connected and semi-simple. We need to show that the algebra $k[G/H]$ is finitely generated over k. Now, the group $G \times T$ acts on G/H by $(g,t)\cdot xH = gxt^{-1}H$ and this action satisfies the finite descendent condition (Exercise 3). Let $\alpha \in \Phi(T,G)$, let U_α be the corresponding one-dimensional subgroup of G, and let x_α: $\mathbf{G}_a \to U_\alpha$ be the canonical isomorphism. Then, for each $t \in T$ and $c \in k$, we have $tx_\alpha(c)t^{-1} = x_\alpha(\alpha(t)c)$. Let $\Psi = \{\alpha, \ldots, \beta\}$. We consider the $B^- \times T$ orbit of $x_\alpha(1) \ldots x_\beta(1)H$, namely, $\{bx_\alpha(\alpha(t)) \ldots x_\beta(\beta(t))H : b \in B^-, t \in T\}$. If the elements of Ψ are linearly independent, then the $\alpha(t), \ldots, \beta(t)$ can be chosen arbitrarily in k^* so the orbit is open; otherwise, the orbit has codimension 1. Thus, in either case we may apply Theorems 16.2, 20.2, and 20.5. QED

Example [40]. Let G be the simply connected simple algebraic group G_2. The six positive roots are α, β, $\alpha + \beta$, $2\alpha + \beta$, $3\alpha + \beta$, $3\alpha + 2\beta$. We shall denote these roots simply by 1, 2, 3, 4, 5, and 6, respectively. Next, we consider sets S of positive roots which are closed under addition. For example, one such subset consists of the roots 236, another of 456. These two subsets are conjugate under the Weyl group, a fact which we indicate by writing {236,456}. The fourteen different conjugacy classes are as follows:

dim = 1: {2, 5, 6}, {1, 3, 4};
dim = 2: {15, 36, 46, 23, 45}, {16, 35, 24}, {26, 56};
dim = 3: {145, 346}, {156, 356, 246}, {236, 456}, {256};
dim = 4: {1456, 2346, 3456}, {2356, 2456};
dim = 5: {13456, 23456};
dim = 6: {123456}.

For those subgroups of dim $\geq$ 3, we may apply Theorems 20.2 and 20.8. Otherwise, the conjugacy classes {2, 5, 6}, {1, 3, 4}, and {16, 35, 24} represent maximal unipotent subgroups in various semi-simple groups. The group {56} is normalized by the semi-simple group corresponding to β and so is covered by Theorem 16.3. This leaves as unaccounted for the class {15, 36, 46, 23, 45}. We should also note that if char $k = 2$ or 3, then there are quasi-closed subsets to be looked at. All of these remaining cases are covered by explicitly showing that there are codimension two embeddings [108]. Therefore, the "Popov-Pommerening conjecture" holds for the group G_2.

Exercises.

1 [28; Section 2.4.2, p.338]. Let $B = TU$ act on an irreducible, quasi-affine variety X. Show that there is an open subset of X which consists of U-separated orbits. (Hint: Use Lemma 3(a) in the appendix to §19 and the fact that orbits of unipotent groups are closed

[10; Proposition 4.10, p.88].)

2. In the proof of Theorem 20.2, show that each p_j appears among the a_i's. (Hint: p_j is a T-weight function so both p_j and $1/p_j$ are everywhere defined on $T{\cdot}y_o$.)

3. In the context of Theorem 20.8, show that the action of the group $G \times T$ on G/H satisfies the finite descendent condition. (Hint: For the T-action, consider $k[G]$ as a subalgebra of $k[U^{-}TU]$; we calculated the T-weights on the latter algebra in the course of proving Theorem 12.1(c).)

§21. Complexity and Finite Generation

A. Statement of results

Our goal in this section is to prove the following theorem of F. Knop.

Theorem 21.1 [61]. *Let G be an affine algebraic group and let $B = TU$ be a Borel subgroup of G (where T is a maximal torus in G and U is a maximal unipotent subgroup of G). Let X be an irreducible, normal (resp. normal and unirational) variety on which G acts morphically. If $c(X) = 0$ (resp. $c(X) = 1$), then $k[X]^U$ is finitely generated over k.*

We first consider some consequences of this theorem.

Corollary 21.2. *Let X and G be as in Theorem 21.1. Then, the algebra $k[X]^G$ is finitely generated over k.*

Proof. First, suppose that G is connected. Then, $k[X]^G = k[X]^B = (k[X]^U)^T$ and we may apply the theorem. In general, let $L = G^{\circ}$. Then, $k[X]^G = (k[X]^L)^{G/L}$. Since $k[X]^L$ is finitely generated over k by what was just proved and since G/L is finite, we see that $k[X]^G$ is finitely generated over k. QED

Corollary 21.3. *Let G be reductive. Let X be an irreducible, normal (resp. normal and unirational) variety on which G acts morphically. If $c(X) = 0$ (resp. $c(X) = 1$), then $k[X]$ is finitely generated over k.*

Proof. We may assume that G is connected and apply Theorems 21.1 and 16.2. QED

Example. We are going to show that the Nagata counter-example gives a normal unirational variety X and an action of a reductive group G so that $c(X) = 2$ and $k[X]^U$ is not finitely generated over k. Thus, the extensions of Theorem 21.1 to varieties of complexity ≥ 2 are not obvious. More precisely, we continue with the discussion of Nagata's counter-example given in §8 and show that the variety $X = G_1/H$ has complexity 2 with respect to the natural action of $G_1 \times D$. First, since H is a subgroup of codimension 3 in U_1, we see that $c(G_1/H) = 3$ with respect to the action of G_1 on G_1/H by left translation. Indeed, the B_1^--orbit of eH has dimension, $\dim B_1^- = \dim G_1 - \dim U_1 = \dim G_1 - (\dim H + 3) = \dim X - 3$.

Lemma 21.4. *Suppose char $k \neq 3$. Let $G_1 \times D$ act on $X = G_1/H$ by $(g,d){\cdot}xH =$*

$gxd^{-1}H$. Then $c(X) = 2$.

Proof. Let $y =$

$$\begin{pmatrix} 1 & 1 \\ 0 & 1 \end{pmatrix} \times \cdots \times \begin{pmatrix} 1 & 1 \\ 0 & 1 \end{pmatrix}.$$

A calculation shows that the stabilizer in $B_1^- \times D$ of yH is $\{(1,1),(-1,-1)\}$. Then, $\dim(B_1^- \times D)\cdot y = \dim B_1^- + 1 = \dim G_1 - \dim U_1 + 1 = \dim G_1 - \dim H - 3 + 1 = \dim X - 2$. We now apply Lemma 19.7(a). QED

The lemma gives a normal, unirational variety $X = G_1/H$ on which a reductive group, $G_1 \times D$, acts in such a way that $c(X) = 2$ but $k[X]$ is not finitely generated over k. In constructing the counter-example, we required the elements of H to satisfy three equations. If there were only one or two equations, then the variety X would have complexity ≤ 1 with respect to the action of $G_1 \times D$ defined above and $k[X]$ would be finitely generated over k by Corollary 21.3.

B. Proof of Theorem 21.1

Our main tools here will be Gordan's lemma, a famous and very useful result in nineteenth century invariant theory, and the theory of divisors. (For a synopsis of the latter, see the Appendix to §4.)

Theorem 21.5 (Gordan's lemma). *For each system of finitely many homogeneous linear inequalities with integer coefficients, there exists a finite set of non-negative integer solutions so that all other non-negative integer solutions are linear combination of these with non-negative integer coefficients.*

Let us restate this theorem symbolically. Suppose that we are given a system of m homogeneous inequalities, say $\Sigma_{j=1}^{n} a_{ij}x_j \geq 0$, where each $a_{ij} \in \mathbf{Z}$. Then, there are finitely many solutions, say $(x_{1i},\ldots,x_{ni})$ for $i = 1,\ldots,r$ such that (i) each x_{hi} is a non-negative integer; (ii) if $(y_1,\ldots,y_n)$ is any solution such that each y_i is a non-negative integer, then there are non-negative integers $c_1,\ldots,c_r$ so that $(y_1,\ldots,y_n) = c_1(x_{11},\ldots,x_{n1}) + \ldots + c_r(x_{1r},\ldots,x_{nr})$.

An elementary combinatoric proof of Gordan's lemma may be found in [65; Proposition 6.3]. However, from the standpoint of these notes and as is made clear in [105; Section 1.4], Gordan's lemma is really about finite generation for invariants of tori and that is how we shall prove it.

Proof. We first assume that each equation is an equality. Indeed, if the ith equation is $\Sigma_{j=1}^{n} a_{ij}x_j \geq 0$, we add a variable z_i and change the equation to $\Sigma_{j=1}^{n} a_{ij}x_j - z_i = 0$. The solutions to the new system are in one-to-one correspondence with the solutions to the old system. Next, let D be an m-dimensional torus; we denote elements in D by $t = (t_1, \ldots, t_m)$. Let χ_i be the character of D given by $\chi_i(t_1, \ldots, t_m) = t_i$. Let $R = k[x_1, \ldots, x_n]$ be a polynomial algebra and let μ_i, $1 \leq i \leq n$, be characters of D given by $\mu_j = a_{1j}\chi_1 + \ldots + a_{mj}\chi_m$. We define an action of D on R by $t\cdot x_i = \mu_i(t)x_i$. A monomial in

R, $x_1^{d(1)} \dots x_n^{d(n)}$, is fixed by D if and only if $a_{11}d(1) + \dots + a_{1n}d(n) = 0, \dots, a_{m1}d(1) + \dots + a_{mn}d(n) = 0$. Thus, the monomials fixed by T correspond to the solutions of the system of equations. But there are finitely many T-invariant monomials which generate the algebra of T-invariants (Theorem A); the exponents of these generators have the desired properties. QED

• *Let X be an irreducible, normal variety on which G acts morphically. If $c(X) = 0$, then $k[X]^U$ is finitely generated over k.*

Proof. Since $c(X) = 0$, there is an $x_0 \in X$ such that the orbit $B{\cdot}x_0$ is open in X (Lemma 19.7(a)). Let D be the (finite) set consisting of all prime divisors in $X - B{\cdot}x_0$, i.e., the set of all closed, irreducible subvarieties Z in $X - B{\cdot}x_0$ such that $\dim Z = \dim X - 1$. To each element Z in D is associated a discrete valuation of $k(X)$ which we shall denote by ν_Z. We note that an element $f \in k(X)$, which is regular on $B{\cdot}x_0$, is in $k[X]$ if and only is $\nu_Z(f) \geq 0$ for all Z in D. Indeed, if f is not everywhere defined on $k[X]$, then f has a pole since X is normal (Theorem 7(b), Appendix, §4).

Let $\chi_1,\dots,\chi_p$ be a basis of $\Gamma(X)$ over $\mathbf{Z}$. For each $i = 1,\dots,p$, let f_i be an element in $k(X)$ such that $b{\cdot}f_i = \chi_i(b)f_i$ for all $b \in B$. Since $B{\cdot}x_0$ is open in X, each f_i and f_i^{-1} is defined on $B{\cdot}x_0$. Let $\chi \in \Gamma(X)$, say $\chi = e(1)\chi_1 + \dots + e(p)\chi_p$. The function $f_\chi = f_1^{e(1)} \dots f_p^{e(p)}$ satisfies $b{\cdot}f_\chi = \chi(b)f_\chi$ for all $b \in B$. Furthermore, if f' is any other element in $k(X)$ with $b{\cdot}f' = \chi(b)f'$ for all $b \in B$, then $f' = cf_\chi$ for some $c \in k$; indeed, $f'/f_\chi \in k(X)^B = k$.

Let $M = \{\chi \in \Gamma(X) : \nu_Z(f_\chi) \geq 0 \text{ for all } Z \in D\}$. If $\chi = e(1)\chi_1 + \dots + e(p)\chi_p$, then $\nu_Z(f_\chi) \geq 0$ for all $Z \in D$ if and only if $e(1)\nu_Z(f_1) + \dots + e(p)\nu_Z(f_p) \geq 0$. (We may assume in what follows that each $e_i \geq 0$. Indeed, if $e_i < 0$ we replace f_i by f_i^{-1} and proceed as below.) Applying Gordan's Lemma to the system of inequalities $e(1)\nu_Z(f_1) + \dots + e(p)\nu_Z(f_p) \geq 0$, $Z \in D$, we see that there are finitely many solutions, say $v_i = (e_{i1},\dots,e_{ip})$ for $i = 1,\dots,m$, such that (i) each e_{ij} is a non-negative integer and (ii) if $y = (y_1,\dots,y_p)$ is a solution in non-negative integers, then there are non-negative integers $c_1,\dots,c_m$ such that $y = \Sigma_i c_i v_i$. For each $i = 1,\dots,m$, let

$$F_i = f_1^{e_{i1}} \dots f_p^{e_{ip}}.$$

Each F_i is in $k[X]$ since each f_j (and f_j^{-1}) is regular on $B{\cdot}x_0$ and $\nu_Z(F_i) \geq 0$ for all Z in D. Since each $f_j \in k(X)^U$, we see that $F_i \in k[X]^U$.

Let $F_1,\dots,F_m,\dots,F_r$ be all the F's obtained in this way. We now show that $k[X]^U = k[F_1,\dots,F_r]$. Let $f \in k[X]^U$; we may assume that f is a T-weight vector, say $t{\cdot}f = \chi(t)f$ for all $t \in T$. Since $f_\chi = cf$ for some $c \in k$ and since $f \in k[X]$, we see that $\chi \in M$. If $\chi = e_1\chi_1 + \dots + e_p\chi_p$ (where, as usual, we assume that each $e_i \geq 0$), then there are non-negative integers $c_1,\dots,c_m$ so that $(e_1,\dots,e_p) = \Sigma c_i(e_{i1},\dots,e_{ip})$. From this we see that

$$f_\chi = f_1^{e_1} \dots f_p^{e_p} = F_1^{c_1} \dots F_m^{c_m}.$$

QED

• *Let X be a normal, irreducible unirational variety on which G acts morphically. If $c(X) = 1$, then $k[X]^U$ is finitely generated over k.*

We shall divide this proof into five steps. Before starting it, let us introduce some notation. By assumption X is unirational with $c(X) = 1$. We apply Lemma 20.3 to see that there is a $t \in k(X)^B$ so that $k(t) = k(X)^B$. Next, let $\Gamma(X)$ have $\{\chi_1,\ldots,\chi_p\}$ as a $\mathbf{Z}$-basis. For each $i = 1,\ldots,p$, let f_i be an element in $k(X)$ so that $b{\cdot}f_i = \chi_i(b)f_i$ for all $b \in B$. For any $\chi \in \Gamma(X)$, say $\chi = e(1)\chi_1 + \ldots + e(p)\chi_p$, we set $f_\chi = f_1^{e(1)} \ldots f_p^{e(p)}$; if $b \in B$, then $b{\cdot}f_\chi = \chi(b)f_\chi$.

(1) *There is an open, non-empty, B-stable subset of X, say X_o, on which t is defined and so that $t\colon X_o \to S \subset \mathbf{A}^1$ is a quotient map for the action of B. In particular, $k[X_o]^B = k[S]$. Furthermore, we may assume that f_χ and f_χ^{-1} are in $k[X_o]$ for all $\chi \in \Gamma(X)$.*
Proof. By Theorem 19.5, there is a non-empty, B-stable open subset X' of X such that the quotient X'/B exists. Let $\pi\colon X' \to X'/B$ be the canonical map. Since π is a quotient mapping, it is separable and $k(X'/B) = k(X')^B$ (Lemma 19.4). Hence, $k(X'/B) = k(X')^B = k(X)^B = k(t)$. The function t is defined on an open normal subset Y of X'/B. Since $t\colon Y \to \mathbf{A}^1$ is a birational map, it defines an isomorphism between open subsets Y' of Y and $S = t(Y')$ of $\mathbf{A}^1$. Let $X_o = \pi^{-1}(Y')$. The mapping t (or more precisely, $t \circ \pi$) from X_o to S is separable and separates B-orbits. Since S is normal, t is a quotient map on X_o (preliminary (2), Appendix, §19). In particular, $k[X_o]^B = k[S]$. Since $b{\cdot}f_i = \chi_i(b)f_i$ for all $b \in B$, the zeros and poles of each f_i are B-stable. We remove all of them from X_o to see that each f_i and f_i^{-1} is defined on the resulting open subset, which we rename X_o, and whose image in X'/B is open by preliminary (3), Appendix, §19. QED

Let D be the set of all prime divisors of X which are contained in $X - X_o$. To each Z in D is associated a discrete valuation ν_Z of $k(X)$. We note that an element $f \in k(X)$ is in $k[X]$ if and only if $f \in k[X_o]$ and $\nu_Z(f) \geq 0$ for all $Z \in D$ (Theorem 7, Appendix to §4).

(2) *As a vector space, $k[X]^U$ is spanned by all elements of the form hf_χ where $h \in k[S]$ and $\nu_Z(hf_\chi) \geq 0$ for all $Z \in D$.*
Proof. We first note that each hf_χ is in $k(X)^U$ since both h and f_χ are. Then hf_χ is defined on X_o and, by the divisor condition, on $X - X_o$. Now, let $f \in k[X]^U$. We may assume that there is a $\chi \in X(T)$ with $t{\cdot}f = \chi(t)f$ for all $t \in T$. Then, $h = f/f_\chi \in k[X_o]^B = k[S]$ and $f = hf_\chi$. Since $f \in k[X]$, we see that $\nu_Z(hf_\chi) \geq 0$ for all $Z \in D$. QED

Let $Z \in D$. The restriction of ν_Z to $k(X)^B = k(t)$ is a discrete valuation. It is either trivial or has the form $n_Z\nu_{x(Z)}$ where $n_Z \in \mathbf{N}$, $x(Z)$ is a suitable point in $\mathbf{P}^1$, and $\nu_{x(Z)}$ is the discrete valuation of $k(t)$ induced by $x(Z)$. If the restriction is trivial, we take $n_Z = 0$ and $x(Z)$ to be any point in $\mathbf{P}^1$. Thus, for each $Z \in D$ and $h \in k[S] = k[X_o]^B \subset k(t)$, we have $\nu_Z(hf_\chi) = \nu_Z(h) + \nu_Z(f_\chi) = n_Z\nu_{x(Z)}(h) + \nu_Z(f_\chi)$. We identify $\mathbf{A}^1$ (and S) with open subsets of $\mathbf{P}^1$. If any of the $x(Z)$ is in $S = t(X_o)$, we shrink X_o by removing (the B-orbit) $t^{-1}(x(Z))$ from X_o. Thus, we may assume that no $x(Z)$ is in S.

Let $\varphi\colon \mathbf{A}^2 - \{0\} \to \mathbf{P}^1$ be the canonical mapping. Let $k(x,y)$ be the field of rational functions on $\mathbf{A}^2$. We recall that an element $f(x,y)/g(x,y)$ in $k(x,y)$ is said to be *homogeneous of degree r* if $f(x,y)$ (resp. $g(x,y)$) is homogeneous of degree p (resp. q) and $p - q = r$. Then $k(\mathbf{P}^1)$ may be identified with the rational functions in $k(x,y)$ which are homogeneous of degree 0.

The set $S_1 = \varphi^{-1}(S)$ is open in $\mathbf{A}^2$. Let $L(Z) = \varphi^{-1}(x(Z))$. We note that $L(Z)$

induces a discrete valuation $v_{L(Z)}$ of $k(\mathbf{A}^2)$ whose restriction to $k(\mathbf{P}^1)$ is $v_{x(Z)}$. Let $k[S_1]_{\text{hom}}$ denote the set of everywhere defined homogeneous rational functions on S_1.

Let $k[\Gamma]$ be the group algebra of $\Gamma(X)$ where the operation on $\Gamma(X)$ will be denoted by multiplication. Let $R_1 \subset k(x,y) \otimes k[\Gamma]$ be the subalgebra which as a vector space is generated by all elements $h \otimes \chi$ where $h \in k[S_1]_{\text{hom}}$, $\chi \in \Gamma(X)$, and $n_Z v_{L(Z)}(h) + v_Z(f_\chi) \geq 0$ for all $Z \in D$. The torus k^* acts on $k(x,y)$ by $c{\cdot}x = cx$, $c{\cdot}y = cy$.

(3) *The algebra C, consisting of all k^*-invariants on R_1, is $k[X]^U$.*
Proof. Each element in C is a sum of elements $h \otimes \chi$ where $h \in k[S_1]$ is a rational function homogeneous of degree 0 (which may be identified with an element in $k[S]$). The map $h \otimes \chi \to hf_\chi$ gives an isomorphism from C onto $k[X]^U$ by (2). (The map is injective since the hf_χ have different T-weights.) QED

(4) *R_1 is a finitely generated k-algebra.*
Proof. The complement of S_1 in $\mathbf{A}^2$ contains finitely many lines $L(i)$ each of which is the null space of a linear functional on $\mathbf{A}^2$, say $t_i = a_i x + b_i y$. Since no $x(Z)$ is in S, for each $Z \in D$ there is an $i(Z)$ so that $v_{L(Z)} = v_{L(i(Z))}$ on $k(\mathbf{P}^1)$.

Let M be the set of all elements in R_1 having the form $\Pi_i t_i^{v(i)} \otimes \chi$ where $v(i) \in \mathbf{Z}$ and $n_Z v(i(Z)) + v_Z(f_\chi) \geq 0$ for all $Z \in D$. We show now that R_1 is generated by $x \otimes 1$, $y \otimes 1$, and M. Indeed, let $h \in k[S_1]_{\text{hom}}$. Then h is a homogeneous function in $k(x,y)$ whose only possible poles are at the $L(i)$. Thus, $h = p\Pi t_i^{v(i)}$ where $p \in k[x,y]$ is not divisible by any t_i. Furthermore, $n_Z v_{L(Z)}(h) = n_Z v(i(Z))$.

Finally, if $f_\chi = f_1^{e(1)} \ldots f_p^{e(p)}$, then the conditions $n_Z v(i(Z)) + v_Z(f_\chi) \geq 0$ for all $Z \in D$ mean that $n_Z v(i(Z)) + e(1)v_Z(f_1) + \ldots + e(p)v_Z(f_p) \geq 0$ for all $Z \in D$. Thus, by Gordan's lemma (modified perhaps as in the proof where $c(X) = 0$ to include the case where some of the v's and e's may be negative), M is a finitely generated monoid and, then, R_1 is finitely generated. QED

(5) *$k[X]^U$ is finitely generated over k.*
Proof. This follows from (3), (4), and Theorem A. QED

§22. Spherical Subgroups

Let G be a connected, reductive group. Let H be a spherical subgroup of G, i.e., any Borel subgroup of G has an open orbit on G/H. Such subgroups have arisen in a number of examples we have encountered. In this section, we make a more systematic study of their basic properties. Most of these follow very quickly from the machinery we have established in previous sections.

Theorem 22.1. *Let G be reductive and let H be a spherical subgroup of G. Let A be any finitely generated, commutative k-algebra on which G acts rationally. Then A^H is finitely generated over k.*
Proof. Since $c(G/H) = 0$, the algebra $k[G/H]$ is finitely generated over k (Corollary 21.3). We now apply Theorem 9.3. QED

Theorem 22.2. *Let G be simply connected and semi-simple and let U be a maximal unipotent subgroup of G. Let H be a connected, spherical subgroup of G such that X(H) = {0}. Then* $^U k[G/H]$ *is a polynomial algebra.*

Proof. Applying Theorem 20.6 and Lemma 20.1, we see that $k[G/H]$ is a unique factorization domain. Also, since $X(H) = \{0\}$, H is an observable subgroup of G (Corollary 1.5) and we may apply Theorem 20.2 to $X = G/H$. QED

Let X be an irreducible affine variety on which G acts morphically and let $A = k[X]$. We construct the algebra $R = (A^U \otimes k[G/U^-])^T$ as in §15. The algebra R is finitely generated over k (Lemma 16.5); let grX denote the corresponding affine variety. We know that dim grX = dim X (Corollary 16.7) and that $^U k[X]$ and $^U k[\mathrm{gr}X]$ are isomorphic as T-algebras (Lemma 15.3).

Theorem 22.3 [89; Theorem 8, p.331]. *Let X be an irreducible affine variety on which G acts morphically. The following statements are equivalent.*

(1) *The action of G on k[X] is multiplicity-free;*
(2) $c(X) = 0$;
(3) *any Borel subgroup of G has an open orbit in X;*
(4) *there is an element* $x \in X$ *such that the orbit* $G{\cdot}x$ *is open in X and* G_x *is spherical;*
(5) *the action of G on k[grX] is multiplicity-free;*
(6) $c(\mathrm{gr}X) = 0$;
(7) *any Borel subgroup of G has an open orbit in grX;*
(8) *grX is an S-variety.*

Proof. Let $B = TU$ be a Borel subgroup of G. According to Lemma 19.7, statements (1), (2), and (3) are equivalent. Similarly, statements (5), (6), and (7) are equivalent. Now, suppose that (3) holds and that the orbit $B{\cdot}x$ is open in X. Then, $G{\cdot}x$ is open in X and G_x is spherical since $B{\cdot}x$ is open in $G{\cdot}x$. Hence (4) is true. If statement (4) holds, then the action of B on $k[X]$ is multiplicity-free since $k[X] \subset k[G{\cdot}x] \subset k[G/G_x]$ and the action of G on $k[G/G_x]$ is multiplicity-free (by Lemma 19.7(b)). Hence, statement (1) holds. Since $^U k[X]$ and $^U k[\mathrm{gr}X]$ are isomorphic as T-algebras, we see that (1) and (5) are equivalent. If statement (7) holds, then G has an open orbit on grX and grX is an S-variety by Corollary 15.7(b). Finally, suppose that grX is an S-variety, say, grX is the closure of the orbit $G{\cdot}w$ where G_w contains U. Then, the action of B on $k[\mathrm{gr}X]$ is multiplicity-free by Lemma 17.1(b) since $k[\mathrm{gr}X] \subset k[G/G_w]$; thus, statement (5) holds. QED

The tools are also in place to study affine embeddings of the homogeneous space G/H.

Lemma 22.4. *Let X and* $x \in X$ *satisfy the (equivalent) conditions of Theorem 22.3 and let* $w \in \mathrm{gr}X$ *be such that the orbit* $G{\cdot}w$ *is open in grX and* G_w *contains a maximal unipotent subgroup of G. If* dim (grX - $G{\cdot}w$) $\leq$ dim grX - 2, *then* dim (X - $G{\cdot}x$) $\leq$ dim X - 2.

Proof. Let Y be an irreducible component of the boundary of X - $G{\cdot}x$ and let I be the corresponding ideal in $k[X]$. According to Corollary 15.7 and Exercise 2, §15, dim Y = dim $k[X]/I$ = dim $k[\mathrm{gr}X]/I(R)$ $\leq$ dim grX - 2 = dim X - 2. QED

Theorem 22.5. *Let H be a spherical subgroup of G. Let X be an irreducible affine variety on which G acts morphically; suppose that there is a point $x \in X$ such that the orbit $G{\cdot}x$ is open in X and $G_x = H$. Let ${}^Uk[X] = k[f_1, \dots, f_r]$ where f_i is non-zero and has T-weight ω_i. Let $\mathbf{Z}E = \{\Sigma_i c_i\omega_i : c_i \in \mathbf{Z}\}$. Then, $\dim (X - G{\cdot}x) \leq \dim X - 2$ if $\mathbf{Z}E \cap X^+T \subset \mathbf{Q}_+E$.*

Proof. We know that ${}^Uk[X] = {}^Uk[\mathrm{gr}X]$ are isomorphic as T-algebras and that gr$X$ is an S-variety (Theorem 22.3), say gr$X$ = cl($G{\cdot}w$). Thus, we may apply Theorem 17.5 to see that $\dim (\mathrm{gr}X - G{\cdot}w) \leq \dim \mathrm{gr}X - 2$. The theorem now follows from Lemma 22.4. QED

Note. From the point of view taken in these notes, Theorem 4.3 shows the importance in invariant theory of studying embeddings of homogeneous spaces. The particular case of spherical subgroups has been studied extensively, e.g., see the survey article [60]. Our approach has been very much influenced by [80].

Example. We shall show that the converse to Theorem 22.5 does not hold in general. In so doing, we follow the presentation in [80] and use M. Krämer's table [64; p.149]. Let char $k = 0$. The subgroup $H = \mathrm{Spin}_7 \times \mathrm{SO}_2$ is spherical in $G = \mathrm{SO}_{10}$ and $X = G/H$ is affine according to Corollary 4.6. Let $\omega_1, \dots, \omega_5$ be the fundamental weights of G. Krämer proved that the irreducible representations of G having a non-zero vector fixed by H are precisely those having highest weights in the semi-group generated by

$$e_1\omega_1 + e_2\omega_2 + e_4\omega_4 + e_5\omega_5$$

where $2e_1 \geq |e_4 - e_5|$ and $2e_1 - (e_4 - e_5) \equiv 0 \pmod 4$. Thus, $2\omega_1$ and $\omega_1 + 2\omega_4$ are highest weights on $k[G/H]$ and, so, $4\omega_4 \in \mathbf{Z}E \cap X^+(T)$ in the terminology of Theorem 22.5. But $4\omega_4$ is not in $\mathbf{Q}_+E$.

Theorem 22.6. *Let G be connected, reductive and let H be a spherical subgroup of G. Each Borel subgroup of G has only finitely many orbits on G/H.*

Proof. (We follow an argument of M. Brion [13].) According to Lemma 1.4, there is a finite-dimensional G-module V and a vector $v \in V$ such that $H = \{g \in G : g{\cdot}v \in kv\}$ and $h{\cdot}v = \chi(h)v$ for all $h \subset H$, where $\chi \in X(H)$. Let $G_1 = G \times k^*$ and $B_1 = B \times k^*$, where B is a Borel subgroup of G. The group G_1 acts on V by $(g,c){\cdot}w = cg{\cdot}w$ for all $w \in V$. The stabilizer in G_1 of v is $H_1 = \{(h, \chi(h)^{-1}) : h \in H\}$ and H_1 is spherical in G_1. Indeed, the projection mapping from G_1 to G gives a morphism from G_1/H_1 to G/H and the inverse image of a B-orbit is a B_1-orbit. Also, if B_1 has finitely many orbits on G_1/H_1, then B has finitely many orbits on G/H. Thus, we may assume from now on that H is observable in G.

Let V be a finite-dimensional G-module and let $v \in V$ be such that $G_v = H$ and $G/H \cong G{\cdot}v$ (Theorem 1.12). Let X_1 be the closure of the orbit $G{\cdot}v$ and let $A = k[X_1]$. We construct gr$A$, R, and D as in §15 and let Y,Z, and W be the corresponding affine varieties (Lemma 16.5). Then, Y is an S-variety since Z is an S-variety by Theorem 22.3 and R is integral over gr$A$ by Theorem 15.5. Also, $\dim Y = \dim X_1$ (Corollary 16.7). According to Theorem 15.14, there is a k^*-equivariant, flat mapping $p: W \to \mathbf{A}^1$ such that $p^{-1}(0) = Y$ and $p^{-1}(u) = X_1$ for $u \neq 0$; in particular, we may consider X and

Y to be subsets of W. Let W' be the G-invariant subset of W consisting of those points having G-orbits of the highest possible dimension. Then W' is open in W (Lemma 19.3) and is irreducible since W is; the fibers of $p \mid W'$ are G-orbits and, in fact, $p^{-1}(u) = G/H$ if $u \neq 0$. By choosing one such fiber, we may consider G/H to be a subset of W'.

Now, for s a non-negative integer, let $X^{(s)}$ be the set consisting of all points $x \in X = G/H$ such that $\dim B{\cdot}x \leq s$; each of the $X^{(s)}$ is closed and is B-invariant (Lemma 19.3). Suppose that G/H has infinitely many B-orbits of dimension r. Then, $\dim X^{(r)} \geq r + 1$. We replace $X^{(r)}$ by one of its connected components, if necessary, and assume from now on that it is irreducible. Let E be the closure of $k^*{\cdot}X^{(r)}$ in W'; then E is B-invariant. Also, $G{\cdot}E = W'$ since $G{\cdot}E$ is k^*-invariant, closed in W' (Theorem 5.3) and p is k^*-equivariant. It follows that E contains a point in the fiber of p above every $u \in \mathbf{A}^1$. We now consider $p \mid E$ and let $X_0^{(r)}$ be the fiber above $\{0\}$. Then $\dim X_0^{(r)} \geq \dim X^{(r)} \geq r + 1$ since $X^{(r)}$ is the fiber in $k^*{\cdot}X^{(r)}$ above any non-zero $u \in k^*$. Also, $\dim B{\cdot}x \leq r$ for all $x \in X_0^{(r)}$ since B-orbits on $k^*{\cdot}X^{(r)}$ have dimension $\leq r$. It follows that the fiber over $\{0\}$ contains infinitely many B-orbits. But, since Y is an S-variety, the fiber over $\{0\}$ is contained in the G-orbit of a point, say $G{\cdot}y$, whose stabilizer contains a maximal unipotent subgroup; the Bruhat decomposition shows that there are only finitely many B-orbits on $G{\cdot}y$. QED

Exercise.

1 (M. Brion). Let G be a connected, linear algebraic group and let H be a closed subgroup of G. Show that $k[G/H]$ is multiplicity-free if and only if H'' is a spherical subgroup of G.

§23. Finite Generation of Induced Modules

A. Condition (FM)

In this section, we shall consider the following question: suppose that $k[G]^H$ is finitely generated over k and that E is a finite-dimensional H-module. Is $\mathrm{ind}_H{}^G E$ a finitely generated $k[G]^H$-module? To facilitate the terminology, we make the following

Definition. Let H be a closed subgroup of G. We say that H satisfies *condition (FM) in G* if (i) $k[G]^H$ is finitely generated over k and (ii) whenever E is any finite-dimensional H-module, then $\mathrm{ind}_H{}^G E$ is a finitely generated $k[G]^H$-module.

Lemma 23.1. *Let H be a closed subgroup of G.*
(a) *Let L be a closed subgroup of G such that* $H \subset L \subset H''$. *If H satisfies condition (FM) in G, then so does L.*
(b) *Let L be a closed subgroup of G such that H is a subgroup in L of finite index. Then H satisfies condition (FM) in G if and only if L does.*
(c) *H satisfies condition (FM) in G if and only if* H° *satisfies condition (FM) in* G°.
(d) *If* $k[G]^H$ *is finitely generated over k and if* $\mathrm{ind}_H{}^G E$ *is a finitely generated* $k[G]^H$*-module for all irreducible, finite dimensional H-modules E, then H satisfies condition (FM) in G.*
(e) *Let B(H) be a Borel subgroup of H. Then H satisfies condition (FM) in G if and only*

if $B(H)$ does.

(f) *Let R be a reductive group which is contained in the normalizer of H in G. Let $U(R)$ be a maximal unipotent subgroup of R. Then H satisfies condition (FM) in G if and only if $HU(R)$ does.*

(g) *If $k[G]^H$ is finitely generated over k and if the finite-dimensional H-module E is contained in a finite-dimensional G-module, then $\mathrm{ind}_H{}^G E$ is a finitely generated $k[G]^H$-module.*

(h) *Let H be an observable subgroup of G such that $k[G]^H$ is finitely generated over k. Then H satisfies condition (FM) in G.*

Proof. To prove statement (a), we first use the properties of the "priming correspondence" given in Lemma 1.1 to see that $L' = H'$, i.e. $k[G]^L = k[G]^H$. Now let E be any finite-dimensional L-module. Then $\mathrm{ind}_L{}^G E = (k[G] \otimes E)^L \subset (k[G] \otimes E)^H$. By assumption, the last space is a finitely generated $k[G]^H$-module; since $k[G]^H$ is finitely generated over k, $\mathrm{ind}_L{}^G E$ is also a finitely generated $k[G]^H$-module.

To prove (b), we first note that $k[G]^L$ is finitely generated over k if and only if $k[G]^H$ is (Theorem 4.1). Now let E be a finite-dimensional H-module and suppose that L satisfies condition (FM) in G. We shall show that $\mathrm{ind}_H{}^G E$ is a finitely generated $k[G]^H$-module. Let E_1 be a finite-dimensional L-module which contains E (Lemma 1.9). It suffices to show that $\mathrm{ind}_H{}^G E_1$ is a finitely generated $k[G]^H$-module (since it contains $\mathrm{ind}_H{}^G E$ as a $k[G]^H$-submodule). But $\mathrm{ind}_H{}^L E_1 = k[L/H] \otimes E_1$ by (P6),§6. Since $k[L/H]$ is finite-dimensional, $\mathrm{ind}_H{}^L E_1$ is finite-dimensional. By assumption, $\mathrm{ind}_L{}^G(\mathrm{ind}_H{}^L E_1) = \mathrm{ind}_H{}^G E_1$ is a finitely generated $k[G]^L$-module. Since $k[G]^L \subset k[G]^H$, it is a finitely generated $k[G]^H$-module.

Conversely, let E be a finite-dimensional L-module and suppose that H satisfies condition (FM) in G. Then $\mathrm{ind}_L{}^G E = (k[G] \otimes E)^L \subset (k[G] \otimes E)^H$. The latter module is finitely generated over $k[G]^H$ by assumption. But $k[G]^H$ being finitely generated and integral over $k[G]^L$ is a finitely generated $k[G]^L$-module. Thus, $\mathrm{ind}_L{}^G E$ is a finitely generated $k[G]^L$-module.

According to statement (b), the subgroup H satisfies condition (FM) in G if and only if H° does. Thus, in proving (c), we may assume that H is connected. Let $G = \cup_i g_i G^\circ$ be the coset decomposition of G with respect to G°, where $g_1 = e$ is the identity in G. Then, $k[G] = \oplus_i k[g_i G^\circ]$ and, since H is connected, we have $k[G/H] = \oplus_i k[g_i G^\circ]^H = \oplus_i k[G^\circ/H]$. Let E be an H-module. Then, $\mathrm{ind}_H{}^G E = (k[G] \otimes E)^H = (\oplus_i k[g_i G^\circ] \otimes E)^H = \oplus_i (k[g_i G^\circ] \otimes E)^H$. Since the first term in this direct sum $(k[G^\circ] \otimes E)^H = \mathrm{ind}_H{}^{G^\circ} E$, statement (c) follows immediately.

We prove statement (d) by induction on $\dim E$, using left exactness, (P9), §6. Indeed, let E be a finite-dimensional H-module and suppose that $\mathrm{ind}_H{}^G E'$ is a finitely generated $k[G]^H$-module whenever $\dim E' < \dim E$. If E is irreducible, we are finished by assumption. Otherwise, let E_1 be a non-zero, proper, H-invariant subspace of E. The exact sequence of $k[G]^H$-modules, $0 \to \mathrm{ind}_H{}^G E_1 \to \mathrm{ind}_H{}^G E \to \mathrm{ind}_H{}^G E/E_1$ shows that $\mathrm{ind}_H{}^G E$ is a finitely generated $k[G]^H$-module (since both $\mathrm{ind}_H{}^G E_1$ and $\mathrm{ind}_H{}^G E/E_1$ are).

To prove (e), we may assume by statement (c) that both G and H are connected. Now suppose that $B(H)$ satisfies condition (FM) in G. Then so does H by statement (a) since $B(H)' = H'$ (by Lemma 1.1,(vi)). Conversely, let H satisfy condition (FM) in G and let E be a finite-dimensional $B(H)$-module. Using (d), we may assume that $\dim E = 1$. Then $E_1 = \mathrm{ind}_{B(H)}{}^H E$ is finite-dimensional by Exercise 1, §12; hence, $\mathrm{ind}_{B(H)}{}^G E =$

$\mathrm{ind}_H{}^G E_1$ is a finitely generated $k[G]^H$-module.

To prove (f), we may assume that R is connected. Furthermore, we may assume that H is also connected. Indeed, we apply statement (b) and use the fact that $H°U(R)$ is the connected component of the identity in $HU(R)$. Next, the algebra $k[G]^H$ is finitely generated over k if and only if $k[G]^{HU(R)} = (k[G]^H)^{U(R)}$ is according to Theorem 16.2. Now, let us show that both subgroups H and $HU(R)$ are observable in HR. The subgroup H is normal in HR so H is observable in HR by Theorem 2.1(4). Next, the homomorphism $R \to S = HR/H$ sends $U(R)$ to a unipotent subgroup U_1 of the group S which is observable in S. We may take a representation of S, say on V, and a vector $v \in V$, so that U_1 is the stabilizer of v. The composite mapping $HR \to HR/H = S \to GL(V)$ has $HU(R)$ as the stabilizer of v. Now suppose that H satisfies condition (FM) in G and let E be any finite-dimensional $HU(R)$-module. By Theorem 2.1(7), we may assume that E is an HR-module. Then $(k[G] \otimes E)^{HU(R)}$ is a finitely generated $k[G]^{HU(R)}$-module if and only if $(k[G] \otimes E)^H$ is a finitely generated $k[G]^H$-module (by Theorem 16.8). Conversely, suppose that $HU(R)$ satisfies condition (FM) in G and let E be a finite-dimensional H-module. We may assume that E is an HR-module. Then $\mathrm{ind}_H{}^G E$ is a finitely generated $k[G]^H$-module by Theorem 16.8 as just applied.

To prove statement (g), let E be contained as an H-module in the finite-dimensional G-module E_1. According to (P6), §6, $\mathrm{ind}_H{}^G E_1 = k[G]^H \otimes E_1$ which is a finitely generated (and free) $k[G]^H$-module. Since $k[G]^H$ is finitely generated over k and since $\mathrm{ind}_H{}^G E$ is a $k[G]^H$-submodule of $\mathrm{ind}_H{}^G E_1$, we see that $\mathrm{ind}_H{}^G E$ is a finitely generated $k[G]^H$-module. Finally, statement (h) follows immediately from Theorem 2.1(7) and statement (g). QED

Since Lemma 23.1(h) provides a definitive answer to the question posed above when H is observable, we turn now to subgroups H which are not observable. Here, our answers will not be as complete.

Lemma 23.2. *Let H be a closed subgroup of G. Let χ be a character of H (which may be trivial) such that (i) $L = \{h \in H : \chi(h) = 1\}$ is observable in G and (ii) $k[G/L]$ is finitely generated over k. Then H satisfies condition (FM) in G.*

Proof. Let E be a finite-dimensional H-module and let $M = \mathrm{ind}_L{}^G E = (k[G] \otimes E)^L$. By Lemma 23.1(h), M is a finitely generated $A = k[G/L]$-module. Now, H acts on M and also, by right translation, on A. The action is by the diagonalizable group $H/L = D$. Furthermore, $M^D = (k[G] \otimes E)^H = \mathrm{ind}_H{}^G E$ and $A^D = k[G/H]$. We apply Theorem 16.9 to complete the proof. QED

Lemma 23.3. *Let H be a closed subgroup of G. Suppose that $\mathfrak{R}_u H$ is the unipotent radical of H and that $k[G/\mathfrak{R}_u H]$ is finitely generated over k. Then H satisfies condition (FM) in G.*

Proof. Let χ be any character of H so that $L = \{h \in H : \chi(h) = 1\}$ is observable in G (Corollary 1.5). Then $L/\mathfrak{R}_u H$ is reductive and $k[G/L] = k[G/\mathfrak{R}_u H]^{L/\mathfrak{R}_u H}$ is finitely generated over k by Theorem A. We now apply Lemma 23.2. QED

Theorem 23.4. *Let G be a reductive algebraic group. Let H be a closed subgroup of G such that the variety G/H is irreducible and has complexity ≤ 1. Then H satisfies*

condition (FM) in G.

Proof. Since G/H is irreducible, we see that each coset of G° in G must contain an element of H (Exercise 1). Then, the homogeneous spaces G/H and $G^\circ/(H \cap G^\circ)$ are isomorphic (by preliminary (2), Appendix, §19) since the natural mapping $G^\circ \to G/H$ is surjective, separable and separates $(H \cap G^\circ)$ orbits. Furthermore, the complexity of $G^\circ/(H \cap G^\circ)$ is that of G/H. For if B is a Borel subgroup of G and if BgH is open in G/H with $g \in G^\circ$, then $Bg(H \cap G^\circ)$ is open in $G^\circ/(H \cap G^\circ)$. Applying Lemma 23.1 (b),(c), we may assume from now on that G is connected.

We have seen that $k[G/H]$ is finitely generated over k (Corollary 21.3). According to Lemma 1.4, there is a finite-dimensional G-module V, a vector $v \in V$, and a character $\chi \in X(H)$ such that $H = \{g \in G : g{\cdot}v \in kv\}$ and $h{\cdot}v = \chi(h)v$ for all $h \in H$. Let $H_\chi = \{h \in H : \chi(h) = 1\}$. Since H_χ is the stabilizer in G of v, it is observable in G (Theorem 1.2). We are going to show that the algebra $k[G/H_\chi]$ is finitely generated over k. Then the theorem will follow from Lemma 23.2.

Let $G_1 = G \times k^*$. The group k^* acts on V by scalar multiplication and this (along with the given action of G) gives an action on V by G_1. Let $H_1 = \{g \in G_1 : g{\cdot}v = v\}$. Then $H_1 = \{(h, \chi^{-1}(h)) : h \in H\}$. The group k^* acts on G_1 by $c_1{\cdot}(g,c) = (g,cc_1)$.

The complexity of G_1/H_1 is equal to that of G/H. Indeed, let p: $G_1 \to G$ be given by $p(g,c) = g$. Then p induces a mapping (also denoted by p) from G_1/H_1 to G/H. The fiber over a point gH is just the k^*-orbit of $(g,1)H_1$ and, so, is of dimension 1. Then $B_1 = B \times k^*$ is a Borel subgroup of G_1 and the image under p of the B_1-orbit of $(g,1)H_1$ is the B-orbit of gH. This fact, along with the comment made above on the fibers, shows that $c(G_1/H_1) = c(G/H)$. According to Corollary 21.3, $k[G_1/H_1]$ is finitely generated over k.

The varieties G_1/H_1, G/H_χ, and $G{\cdot}v$ are all normal and the mappings π: $G_1/H_1 \to G{\cdot}v$ and ψ: $G/H_\chi \to G{\cdot}v$ are both bijective. According to Lemma 5.5, $k[G{\cdot}v]$ is finitely generated over k. Also, as in the proof of Lemma 5.5, we see that $k[G/H_\chi]$ is contained in the integral closure R of $k[G{\cdot}v]$ in $k(G/H_\chi)$. Since the mapping $G/H_\chi \to G{\cdot}v$ is injective, $k(G/H_\chi)$ is a finite algebraic extension of $k(G{\cdot}v)$ and R is a finitely generated $k[G{\cdot}v]$ module. Hence, $k[G/H_\chi]$ is finitely generated over k. QED

B. Epimorphic subgroups

We recall that a closed subgroup H of G is said to be epimorphic in G if $H'' = G$ (§1). The next two results show the relationship between epimorphic subgroups and the question posed at the beginning of Part A.

Lemma 23.5. *Let H be a closed subgroup of G. Then H is an epimorphic subgroup of H''.*

Proof. Let $L = H''$ and let $f \in k[L]^H$; it suffices to show that $f \in k$. The function f is contained in a finite-dimensional vector space V of $k[L]$, which is invariant under the action (by right translation) of L. By definition, L is observable in G. Then, by Theorem 2.1(7), V is contained as an L-submodule in a finite-dimensional rational G-module V_1. The stabilizer of f in G is observable (by Theorem 1.12) and contains H. Hence, it also contains L. Thus, $f \in k$. QED

Lemma 23.6. *Let H be a closed subgroup of G and let $L = H''$. Suppose that $k[G]^H$ is finitely generated over k. Let E be a finite-dimensional H-module, If $ind_H{}^L E$ is a finite-dimensional vector space over k, then $ind_H{}^G E$ is a finitely generated $k[G]^H$-module.*
Proof. According to Corollary 1.6, we have $k[G]^L = k[G]^H$. Let $E_1 = \mathrm{ind}_H{}^L E$; then $\mathrm{ind}_L{}^G E_1$ is a finitely generated $k[G]^L$-module by Lemma 23.1(h). Hence, $\mathrm{ind}_H{}^G E = \mathrm{ind}_L{}^G \mathrm{ind}_H{}^L E$ is a finitely generated $k[G]^H$-module. QED

The lemmas indicate the value of studying epimorphic subgroups. The next theorem shows their relationship to the representation theory of *G*.

Lemma 23.7. *For H a closed subgroup of G, the following statements are equivalent.*
(a) *H is epimorphic in G.*
(b) *$k[G/H] = k$.*
(c) *If E is any rational G-module, then $E^H = E^G$.*
(d) *Let V be any rational G-module which is a direct sum of two H-invariant subspaces X and Y; then X and Y are G-invariant.*
(e) *If r and s are homomorphisms of G into a linear algebraic group G^1 such that $r(h) = s(h)$ for all $h \in H$, then $r(g) = s(g)$ for all $g \in G$.*
Proof [5; Théorème 1, p.649]. The equivalence of statements (a), (b), and (c) follows from the definition of an epimorphic subgroup and Corollary 1.6. Now, suppose that (c) holds and let *V* be a *G*-module which is a direct sum of two *H*-invariant subspaces *X* and *Y*. Let $p: V \to X$ be the projection mapping. Since *X* is *H*-invariant, we see that $p \circ h = h \circ p$ for all $h \in H$. Now let $x \in X$; we shall show that $G{\cdot}x \subset X$. Indeed, let $V_1 = \langle G{\cdot}x \rangle$, the finite-dimensional vector space spanned by all $g{\cdot}x$, $g \in G$. Let Z_1 be a finite-dimensional *G*-module containing $p(V_1)$ and consider $p\colon V_1 \to Z_1$. Since $p \circ h = h \circ p$ for all $h \in H$, we have $p \circ g = g \circ p$ for all $g \in G$ by statement (c). Then, $p(g{\cdot}x) = g{\cdot}p(x) = g{\cdot}x$ and, so, $g{\cdot}x \in X$.

Let (d) hold and let r,s be homomorphisms of *G* to G^1 such that $r(h) = s(h)$ for all $h \in H$. Since G^1 is a linear algebraic group, we may assume that $G^1 = GL(V)$, where *V* is a finite-dimensional vector space. The mappings r,s give a homomorphism of *G* into $GL(V \oplus V)$, namely, $g{\cdot}(v_1,v_2) = (r(g)v_1, s(g)v_2)$. Now the vector space $V \oplus V$ is the direct sum of its first factor and the diagonal subspace $Z = \{(v,v) : v \in V\}$. Since $r(h) = s(h)$ for all $h \in H$, both subspaces are *H*-invariant. By assumption, then, *Z* is *G*-invariant. Hence, $r(g) = s(g)$ for all $g \in G$.

To show that statements (e) implies (a), let $f \in k[G]^H$ and let *E* be a finite-dimensional subspace of $k[G]$, invariant under right translation by *G*, and containing *f*. Let $G^1 = E \times G$ with the product defined by $(x,g){\cdot}(x',g') = (x + g{\cdot}x', gg')$. Let r,s be the homomorphisms from *G* to G^1 defined by $r(g) = (0,g)$ and $s(g) = (g{\cdot}f - f, g)$. Then $r = s$ on *H* so, by assumption, $r = s$ on *G*, i.e. $g \cdot f = f$ for all $g \in G$. Thus, $f \in k$.
QED

Definition. An epimorphic subgroup *H* of *G* is said to satisfy *condition (F) in G* if for any finite-dimensional *H*-module *E*, the vector space $\mathrm{ind}_H{}^G E$ is finite-dimensional over *k*. The group *G* is said to satisfy *condition (F)* if each epimorphic subgroup of *G* satisfies condition (F) in *G*.

Obviously, condition (F) is just condition (FM), above, placed in the context of epimorphic subgroups.

Lemma 23.8. *Let H be an epimorphic subgroup of G.*
(a) *Let L be a closed subgroup of G which contains H. Then, L is epimorphic in G and, if H satisfies condition (F) in G, so does L.*
(b) *Let L be a closed subgroup of G such that H is a subgroup in L of finite index. Then H satisfies condition (F) in G if and only if L does.*
(c) *H satisfies condition (F) in G if and only if* H^o *satisfies condition (F) in* G^o.
(d) *If* $\mathrm{ind}_H{}^G E$ *is finite-dimensional for all irreducible, finite-dimensional H-modules E, then H satisfies condition (F) in G.*
(e) *Let* B_H *be a Borel subgroup of H. Then H satisfies condition (F) in G if and only if* B_H *satisfies condition (F) in* G^o.
(f) *Let R be a reductive group which is contained in the normalizer of H in G. Let U(R) be a maximal unipotent subgroup of R. Then H satisfies condition (F) in G if and only if HU(R) does.*
(g) *Let* χ *be a character of H so that* $L = \{h \in H : \chi(h) = 1\}$ *is a Grosshans subgroup of G. Then H satisfies condition (F) in G.*
(h) *Let* $\mathfrak{R}_u H$ *be the unipotent radical of H. If* $k[G/\mathfrak{R}_u H]$ *is finitely generated over k, then H satisfies condition (F) in G.*
Proof. First, let us look at statement (c). We need to show that H° is epimorphic in G°. The subgroup $H \cap G^\circ$ is epimorphic in G° by Exercise 5, §1. Thus, we may replace G by G° and H by $H \cap G^\circ$. The finite group $H^\circ\backslash H$ acts by right translation on the algebra $k[G/H^\circ]$ and its algebra of invariants is $k[G/H]$. Thus, $k[G/H^\circ]$ is integral over $k[G/H]$ since $f \in k[G/H]$ satisfies the equation $\Pi_g(x - g{\cdot}f) = 0$ where g runs over all the elements in $H^\circ\backslash H$. But $k[G/H] = k$ so $k[G/H^\circ] = k$ and H° is epimorphic in G. All of the other statements follow at once from Lemmas 23.1 - 23.3 where we also use Lemma 1.1(vi) for statement (e). QED

Theorem 23.9. *Let G be connected and reductive. Let T be a maximal torus in G, let P be a proper parabolic subgroup of G containing T, and let* $\mathfrak{R}_u P$ *be the unipotent radical of P. If L is any closed subgroup of G which contains* $T\mathfrak{R}_u P$, *then L is epimorphic in G and satisfies condition (F) in G.*
Proof. We first show that $T\mathfrak{R}_u P$ is epimorphic in G. To do this, we may first assume that G is simply connected and then that G is simple. Indeed, if $G = G_1 \times \ldots \times G_r$ and $H = H_1 \times \ldots \times H_r$ is a subgroup of G, then $k[G/H] = \otimes_i k[G_i/H_i]$. Next, let $P = R\mathfrak{R}_u P$ be a Levi decomposition of P where R is a reductive group. Let $U(R)$ be any maximal unipotent subgroup of R which is normalized by T; let $U = U(R)\mathfrak{R}_u P$. Let $\{\alpha_1,\ldots,\alpha_m,\ldots,\alpha_n\}$ be a simple root system in $\Phi(T,G)$ corresponding to U; we shall assume that $\{\alpha_1,\ldots,\alpha_m\}$ is a simple root system in $\Phi(T,R)$ corresponding to $U(R)$. Let $f \in k[G/T\mathfrak{R}_u P]$; we shall show that $f \in k$. Let W be the vector space spanned by all the vectors $u{\cdot}f$ where $u \in U$. If $\dim W = 1$, then TU fixes f and $f \in k$ by Lemma 1.1(vi). Hence, suppose that $\dim W > 1$. We note that each $u{\cdot}f = u_R{\cdot}f$ for some $u_R \in U(R)$ since $\mathfrak{R}_u P \subset G_f$. The vector space W is invariant under TU and the weights of T on W have the form $e_1\alpha_1 + \ldots + e_m\alpha_m$ where each e_i is a non-negative integer (Lemma 3.1). There are non-zero T weight vectors $w \in W$ fixed by U. Let w_o be such a vector and

suppose (to simplify the notation) that w_0 has T-weight $\omega = e_1\alpha_1 + \ldots + e_r\alpha_r$ where $r \leq m$ and each $e_i > 0$. Since U fixes w_0, we have $(\omega,\alpha) \geq 0$ for all roots α of U by Theorem 3.2. If there is a simple root β among $\{\alpha_{r+1},\ldots,\alpha_n\}$ connected to one of the $\alpha_1, \ldots, \alpha_r$, then $(\omega,\beta) < 0$. Thus, $\{\alpha_1,\ldots,\alpha_r\} = \{\alpha_1,\ldots,\alpha_n\}$ and $R = G$, a contradiction. Next, the group $T\mathfrak{R}_uP$ satisfies (F) in G by Theorem 16.4 applied to Lemma 23.8(h). Then L satisfies condition (F) in G by Lemma 23.8(a). QED

Let $B = TU$ be a Borel subgroup of G and let $\omega \in X^+(T)$. In §12, we studied the spaces $E(\omega) = \{f \in k[G]^U : r(t)f = (\iota\omega)(t)f \text{ for all } t \in T\}$. In particular, we showed that the $E(\omega)$ are all of finite-dimension (Theorem 12.1(c)).

Theorem 23.10. *Let G be connected and reductive with Borel subgroup $B = TU$. Let H be an epimorphic subgroup of G and let B_H be a Borel subgroup of H. The following statements are equivalent.*
(a) *H satisfies condition (F) in G;*
(b) *for each $\chi \in X(B_H)$, there are only finitely many $E(\omega)$ containing a non-zero B_H-eigenvector of weight χ.*
Proof. First, H satisfies condition (F) in G if and only if B_H does by Lemma 23.8(e). Thus, we may replace H by B_H from now on. Let ϵ: $k[G] \to k[G]$ be defined by $(\epsilon f)(x) = f(x^{-1})$. Then $\epsilon r_g = \ell_g\epsilon$ for all $g \in G$. Since H is solvable, statement (a) is equivalent to $\text{ind}_H{}^G E$ being finite-dimensional for all H-modules E with $\dim E = 1$ (by Lemma 23.8(d)). Let E be such; let v be a non-zero vector in E and let $\chi \in X(H)$ be chosen so that $h{\cdot}v = \chi(h^{-1})v$ for all $h \in H$. Then $\text{ind}_H{}^G E = (k[G] \otimes E)^H$ may be identified with $E_1 = \{f \in k[G] : r_h f = \chi(h)f \text{ for all } h \in H\}$. Let $F = \epsilon(F_1) = \{f \in k[G] : \ell_{h}f = \chi(h)f$ for all $h \in H\}$. The group G acts by right translation on F. By Theorem 16.10, the vector space F is of finite-dimension if and only if F^U is. Let $\{\omega_i\}$, $i \in I$, be the distinct T-weights on F^U (where T acts by right translation). Then F^U is contained in the direct sum of the $E(\iota\omega_i)$, each of which is finite-dimensional. QED

Corollary 23.11. *Let G be connected and reductive. Let H be a connected, solvable, epimorphic subgroup of G. Let L be a closed subgroup of G which contains H as a normal subgroup and is such that L/H is unipotent. Then H satisfies condition (F) in G if and only if L does.*
Proof. We have seen (Lemma 23.8(a)) that if H satisfies condition (F), then L does too. Suppose now that L satisfies condition (F); we may assume that L is solvable and connected (Lemma 23.8(e)). By induction, we may further assume that L/H is $\mathbf{G}_a$. To show that H satisfies condition (F) in G, we shall prove that statement (b) in Theorem 23.10 holds. Let $\chi \in X(H)$ and let v be a non-zero vector in some $E(\omega)$ such that $h{\cdot}v = \chi(h)v$ for all $h \in H$. We may assume that $\chi \in X(L)$. For $g \in L$, let $g{\cdot}\chi$ be the character of H given by $(g{\cdot}\chi)(h) = \chi(g^{-1}hg)$ for all $h \in H$. Since $\chi \in X(L)$, we see that $g{\cdot}\chi = \chi$ and $h{\cdot}(g{\cdot}v) = (g{\cdot}\chi)(h)g{\cdot}v = \chi(h)g{\cdot}v$ for all $g \in L$, $h \in H$. Let $<L{\cdot}v>$ be the finite-dimensional vector space spanned by all $g{\cdot}v$, $g \in L$. Since L is solvable, there is a non-zero L-eigenvector $w \in <Lv>$ and $g{\cdot}w = \chi(g)w$ for all $g \in L$. Since L satisfies statement (b), Theorem 23.10, so does H. QED

Corollary 23.12. *Let G be connected, reductive and let H be an epimorphic subgroup*

of G which is normalized by a maximal torus T in G. Then H satisfies condition (F) in G if and only if HT does.

Proof. First, we may assume that H is connected (Lemma 23.8(c)). Next, let $B_1 = TV$ be a Borel subgroup in TH where V is the unipotent radical of B_1. We shall show that $B_1 \cap H$ is a Borel subgroup of H. First, the image of H in HT/H is a torus so V is contained in H and $B_1 = T(B_1 \cap H)$. Next, the mapping $H/(B_1 \cap H) \to HT/B_1$ is bijective. Since the variety HT/B_1 is complete and quasi-projective, it is projective [54; Proposition,(f), p.45]. Then, $H/(B_1 \cap H)$ is the normalization of HT/B_1 in $k(H/(B_1 \cap H))$ by Theorem 1.10 and is complete since the normalization of a projective variety is projective [10; AG 18.2]. From this we see that $B_1 \cap H$ is parabolic (and, so, connected) in H; since it is solvable, it is a Borel subgroup of H which is (obviously) normalized by T. Applying Lemma 23.8(e), we may replace H by $B_1 \cap H$ and assume from now on that H is solvable.

If H satisfies condition (F), then so does HT according to Lemma 23.8(a). Suppose, then, that HT satisfies condition (F). We shall show that H satisfies condition (F) by induction on dim HT - dim H. Thus, we may assume that there is a closed, connected, solvable subgroup L of G which is normalized by T, satisfies condition (F) and is such that L/H is $\mathbf{G}_m$. Let $\mu \in X(L)$ be a non-trivial character whose restriction to H is trivial. There are roots α and β in $\Phi(T,H)$ such that $(\mu,\alpha) < 0$ and $(\mu,\beta) > 0$. Indeed, suppose that $(\mu,\alpha) \geq 0$ for all $\alpha \in \Phi(T,H)$. The character μ is conjugate by the Weyl group of T to a unique dominant weight with respect to some given Borel subgroup of G containing T [54; 31.2]. Thus, there is a finite-dimensional G-module V and a vector $v \in V$ so that the stabilizer of v, G_v, contains all of the U_α such that $(\mu,\alpha) \geq 0$. Since G_v is observable and contains H, we see that $G_v = G$ and μ must be trivial.

To show that H satisfies condition (F) in G, we shall show that statement (b) in Theorem 23.10 holds. Let $\chi \in X(H)$ and let v be a non-zero vector in some $E(\omega)$ such that $h{\cdot}v = \chi(h)v$ for all $h \in H$. We may assume that $\chi \in X(L)$. For $g \in L$, let $g{\cdot}\chi$ be the character of H defined by $(g{\cdot}\chi)(h) = \chi(g^{-1}hg)$ for all $h \in H$. Since $\chi \in X(L)$, we see that $g{\cdot}\chi = \chi$ on H and also that $h{\cdot}(g{\cdot}v) = (g{\cdot}\chi)(h)g{\cdot}v = \chi(h)g{\cdot}v$ for all $g \in L$. Let $<L{\cdot}v>$ be the finite-dimensional vector space spanned by all the $g{\cdot}v$, $g \in L$. Since L is solvable, there is a non-zero L-eigenvector $w \in <L{\cdot}v>$ and $g{\cdot}w = \chi'(g)w$ for all $g \in L$ where $\chi' | H = \chi$. Thus, $\chi' = \chi + m\mu$ for some integer m. But only finitely many such m's are possible. For example, suppose that $m < 0$. By Theorem 3.2, $(\chi + m\mu,\beta) \geq 0$ and, so, $m \geq -(\chi,\beta)/(\mu,\beta)$. By induction, $\chi + m\mu$ can be the weight of an L-eigenvector in only finitely many $E(\omega)$. Since L satisfies statement (b) of Theorem 23.10, so does H. QED

Corollary 23.13. *Let G be connected, reductive and let T be a maximal torus in G. Let H be an epimorphic, solvable subgroup of G which contains T. Suppose that there is a root $\beta \in \Phi(T,G)$ so that the unipotent radical of H is normalized by U_β and (ii) there is an $\alpha' \in \Phi(T,H)$ with $(\beta,\alpha') < 0$. Let $L = HU_\beta$. If L satisfies condition (F) in G, then so does H.*

Proof. Let $\chi \in X(H)$ and let $V = E(\omega)$ contain a non-zero H-eigenvector v having weight χ. (We need to show there are only finitely many such ω by Theorem 23.10.) Then $(\chi,\alpha) \geq 0$ for all $\alpha \in \Phi(T,H)$ (Theorem 3.2). Let M be the largest non-negative integer $\leq -(\chi,\alpha')/(\beta,\alpha')$. Let $W = <U_\beta{\cdot}v>$. Then W is L-invariant and contains a non-zero L-eigenvector, say w. The weights of T on W have the form $\chi + m\beta$, where $m \in \mathbf{Z}$ and $m \geq 0$ (Lemma 3.1). If w has weight $\chi + m\beta$, then $(\chi + m\beta,\alpha') \geq 0$, i.e., $m \leq -(\chi,\alpha')/(\beta,\alpha')$. Thus, $E(\omega)$ contains an L- eigenvector of weight $\chi + m\beta$ for some m with $0 \leq m \leq M$. The set of such $E(\omega)$ is finite by Theorem 23.10(b). QED

Note 1. The main sources for this section are [5] and [6] although the particular details of the exposition here (for example, condition (FM) and many of the proofs) are different. Theorem 23.4 was first proved (in an unpublished manuscript) by M. Brion in the case char $k = 0$.

Note 2. There are a number of conditions whose precise relationship is not fully understood. For example, let H be a closed, epimorphic subgroup of G and consider the following:

(F) for any finite-dimensional H-module E, the vector space $\mathrm{ind}_H{}^G E$ is finite-dimensional over k;
(FG) there is a character $\chi \in X(H)$ such that $H_\chi = \{h \in H : \chi(h) = 1\}$ is a Grosshans subgroup of G;
(SFG) the algebra $k[G/\mathfrak{R}_u H]$ is finitely generated over k.

We have seen (in Lemmas 23.2 and 23.3) that (SFG) $\Rightarrow$ (FG) $\Rightarrow$ (F). The Nagata counter-example shows that (F) does not imply (FG) [7; 4.7]. Of course, if the Popov-Pommerening conjecture were to hold, then (SFG) would be true in the case where G is reductive and H is normalized by a maximal torus. It is also known that if G/H has a codimension 2 projective embedding, then (F) holds [6; Théorème 1, p.1343].

F. Bien and A. Borel have conjectured that (F) always holds. And, in fact, if H is an epimorphic subgroup of a reductive group G which is normalized by a maximal torus in G, then (F) does hold [7; Theorem 4.3, p.117].

Exercise.

1. If G/H is irreducible, show that each coset of G° in G contains an element of H. (Hint: by [10; 6.8] the variety G/H is quasi-projective and, in fact, is the orbit $G{\cdot}x$ of a point x in some projective space. If $G = \cup g_j G^\circ$, then $G \cdot x = \cup g_j G^\circ{\cdot}x$. The subvarieties $g_j G^\circ{\cdot}x$ are locally closed in G/H and all have the same dimension.)

References

1. A'Campo - Neuen, A.: Note on a counterexample to Hilbert's fourteenth problem given by P. Roberts. *Indag. Math.* N. S. **5** (3), 253 - 257 (1994)

2. Betten, J., Helisch, W.: Simultaninvarianten bei Systemen zwei- und vierstufiger Tensoren. *Z. angew. Math. Mech.* **75** (10), 753 - 759 (1995)

3. Bialynicki - Birula, A.: On homogeneous affine spaces of linear algebraic groups. *Amer. J. Math.* **85**, 577 - 582 (1963)

4. Bialynicki - Birula, A., Hochschild, G., Mostow, G. D.: Extensions of representations of algebraic linear groups. *Amer. J. Math.* **85**, 131 - 144 (1963)

5. Bien, F., Borel, A.: Sous - groupes épimorphiques de groupes linéaires algébriques I. *C. R. Acad. Sci.* Paris. Ser. I. **315**, 649 - 653 (1992)

6. Bien, F., Borel, A.: Sous - groupes épimorphiques de groupes linéaires algébriques II. *C. R. Acad. Sci.* Paris. Ser. I. **315**, 1341 - 1346 (1992)

7. Bien, F., Borel, A., Kollár, J.: Rationally connected homogeneous spaces. *Invent. math.* **124**, 103 - 127 (1996)

8. Birkes, D.: Orbits of linear algebraic groups. *Annals of Math.* **93**, 459 - 475 (1971)

9. Bogomolov, F. A.: Holomorphic tensors and vector bundles on projective varieties. *Isv. Akad. Nauk SSSR*, Ser. Mat. **42**, No. 6, 1227 - 1287. English translation: *Math USSR, Izv.* **13**, 499 - 555 (1979)

10. Borel, A.: *Linear Algebraic Groups*. Second enlarged edition. Graduate texts in Mathematics **126**. Berlin - Heidelberg - New York: Springer 1991

11. Borel, A., Tits, J.: Groupes réductifs. *Inst. Hautes Études Sci. Publ. Math.* **27**, 55 - 150 (1965)

12. Bourbaki, N.: *Eléments de mathématique*. Vol. XXXIV, Groupes et algébres de Lie, Chapitres 4, 5, et 6. Paris: Hermann 1968

13. Brion, M.: Quelques propriétés des espaces homogènes sphériques. *Manuscripta Math.* **55**, 191 - 198 (1986)

14. Brion, M.: Sur les modules de covariants. *Annales scientifiques de L'École Normale supérieure*, 4th série **26**, 1 - 21 (1993)

15. Brion, M., Luna, D., Vust, Th.: Espaces homogènes sphériques. *Invent. math.* **84**, 617 - 632 (1986)

16. Chevalley, C.: *Séminaire C. Chevalley, Classification des groupes de Lie algébriques*. Paris: Ecole Norm. Sup. 1956 - 58

17. Chevalley, C.: *Fondéments de géométrie algébrique*. Paris: Secrétariat Mathématique 1958

18. Cicurel, R.: Sur les orbites affines des G - variètès admettant une orbite ouverte. *C. R. Acad. Sci.* Paris. **274**, 1316 - 1318 (1972)

19. Cline, E., Parshall, B., Scott, L.: Induced modules and affine quotients. *Math. Ann.* **230**, 1 - 14 (1977)

20. Crapo, H.: Automatic proving of geometric theorems. In: *Invariant Methods in Discrete and Computational Geometry*, 167 - 196. Dordrecht: Kluwer Acad. Publ. 1995

21. Danielewski, W.: Thesis, Warsaw University (1988)

22. Désarménien, J.: An algorithm for the Rota straightening formula. *Discrete Math.* **30**, 51 - 68 (1980)

23. Désarménien, J., Kung, J. P. S., Rota, G.- C.: Invariant theory, Young bitableaux, and combinatorics. *Advances in Math.* **27**, 63 - 92 (1978)

24. Deveney, J., Finston, D.: $\mathbf{G}_a$ actions on $\mathbf{C}^3$ and $\mathbf{C}^7$. *Communications in Algebra* **22** (15), 6295 - 6302 (1994)

25. Dieudonné, J. A., Carrell, J. B.: *Invariant Theory, Old and New*. New York - London: Academic Press 1971

26. Dixmier, J.: Sur les representations unitaires des groupes de Lie nilpotents. IV. *Canadian J. Math.* **11**, 321 - 344 (1959)

27. Dixmier, J.: Solution négative du problème des invariants, d'après Nagata. *Séminaire Bourbaki* 1958 - 9, Expose 175. Paris: Secrétariat Mathématique

28. Dixmier, J., Reynaud, M.: Sur le quotient d'une variété algébrique par un groupe algébrique. Mathematical analysis and applications, Part A. *Advances in Math. Supplementary Studies* **7A**, 327 - 344 (1981)

29. Donkin, S.: *Rational Representations of Algebraic Groups: Tensor Products and Filtrations*. Lecture Notes in Mathematics **1140**. Berlin - Heidelberg - New York: Springer 1985

30. Donkin, S.: Good filtrations of rational modules for reductive groups. *Proc. Symposia in Pure Math.* **47**, 69 - 80 (1987)

31. Donkin, S.: Invariants of unipotent radicals. *Math. Z.* **198**, 117 - 125 (1988)

32. Donkin, S.: The normality of closures of conjugacy classes of matrices.

Invent. math. **101**, 717 - 736 (1990)

33. Elliott, E. B.: *An Introduction to the Algebra of Quantics*. Second edition. Oxford University Press 1913. Reprinted. Bronx, New York: Chelsea Publishing Company 1964

34. Fauntleroy, A.: On Weitzenböck's theorem in positive characteristic. *Proc. Amer. Math. Soc.* **64**, 209 - 213 (1977)

35. Fauntleroy, A.: Algebraic and algebro - geometric interpretations of Weitzenböck's problem. *J. Algebra* **62**, 21 - 38 (1980)

36. Green, J. A.: *Classical Invariants*. Textos de Matemática, Série B. Departamento de Matemática da Universidade de Coimbra 1993

37. Greuel, G - M, Pfister, G.: Geometric quotients of unipotent group actions. *Proc. London Math. Soc.* **67** (3), 75 - 105 (1993)

38. Grosshans, F.: Observable groups and Hilbert's fourteenth problem. *Amer. J. Math.* **95**, 229 - 253 (1973)

39. Grosshans, F.: The invariants of unipotent radicals of parabolic subgroups. *Invent. math.* **73**, 1 - 9 (1983)

40. Grosshans, F.: Hilbert's fourteenth problem for non - reductive groups. *Math. Z.* **193**, 95 - 103 (1986)

41. Grosshans, F.: Constructing invariant polynomials via Tschirnhaus transformations. In: *Invariant Theory*, 95 - 102. Lecture Notes in Mathematics **1278**. Berlin - Heidelberg - New York: Springer 1987

42. Grosshans, F.: Contractions of the actions of reductive groups in arbitrary characteristic. *Invent. math.* **107**, 127 - 133 (1992)

43. Guillemonat, A.: On finite generation of invariants for certain subalgebras of a semisimple Lie algebra. In: *Non - commutative harmonic analysis* (Marseille - Luminy 1976), 77 - 90. Lecture Notes in Mathematics **587**. Berlin - Heidelberg - New York: Springer 1977

44. Hadžiev, Dž.: Some questions in the theory of vector invariants. *Math. USSR Sbornik* **1**, 383 - 396 (1967)

45. Hartshorne, R.: *Algebraic Geometry*. Graduate Texts in Mathematics **52**. Berlin - Heidelberg - New York: Springer 1977

46. Hilbert, D.: *Theory of Algebraic Invariants*. Cambridge Mathematical Library. Cambridge: Cambridge University Press 1993

47. Hochschild, G., Mostow, G. D.: Unipotent groups in invariant theory. *Proc. Nat. Acad. Sci.* USA **70**, 646 - 648 (1973)

48. Horvath, J.: Invariants of certain unipotent groups. Ph. D. dissertation, Yale University 1987.

49. Horvath, J.: On a problem of Pommerening in invariant theory. *J. Algebra* **126**, 300 - 309 (1989)

50. Howe, R.: (GL_n, GL_m) - duality and symmetric plethysm. *Proc. Indian Acad. Sci.* (Math. Sci.) **97**, 85 - 109 (1987)

51. Howe, R.: The first fundamental theorem of invariant theory and spherical subgroups. *Proc. Symposia in Pure Mathematics* **56**, Part 1, 333 - 346 (1994)

52. Howe, R.: Perspectives on invariant theory: Schur duality, multiplicity free actions and beyond. *Israel Math. Conf. Proceedings* **6**. Ramat - Gan, Israel: Gelbart Research Institute for Mathematical Sciences, Emmy Noether Research Institute, Bar Illan University 1995

53. Howe, R., Huang, R.: Projective invariants of four subspaces. *Advances in Math.* **118**, 295 - 336 (1996)

54. Humphreys, J. E.: *Linear Algebraic Groups*. Graduate Texts in Mathematics **21**. Berlin - Heidelberg - New York: Springer 1975

55. Hungerford, T. W.: *Algebra*. New York: Holt, Rinehart and Winston, Inc. 1974

56. Igusa, J. I.: On certain representations of semi - simple algebraic groups and the arithmetic of the corresponding invariants. *Invent. math.* **12**, 12 - 44 (1970)

57. Jantzen, J. C.: *Rational Representations of Algebraic Groups*. Pure and Applied Mathematics **131**. Boston - Orlando: Academic Press 1987

58. Kaplansky, I.: *Commutative Rings*. Boston: Allyn and Bacon, Inc. 1970

59. Klein, F.: *Elementary Mathematics from an Advanced Standpoint, Geometry*. 3rd ed. New York: Dover Publications Inc. 1948. Translated by E. R. Hedrick and C. A. Noble from the third German edition. *Elementarmathematik vom hoheren Standpunkte*, Band 2: Geometrie, The Macmillan Company 1940.

60. Knop, F.: The Luna - Vust theory of spherical embeddings. In: *Proceedings of the Hyderabad Conference on Algebraic Groups* (Ramanan, S. ed). Manoj Prakashan 225 - 249 (1989)

61. Knop, F.: Über Hilberts vierzehntes Problem für Varietäten mit Kompliziertheit eins. *Math. Z.* **213**, 33 - 35 (1993)

62. Kostant, B.: Lie group representations on polynomial rings. *Amer. J. Math.* **85**, 327 - 404 (1963)

63. Kraft, H.: *Geometrische Methoden in der Invariantentheorie*, 2, durchgesehene

Auflage. Braunschweig/Wiesbaden: Friedr. Vieweg & Sohn 1985

64. Krämer, M.: Sphärische Untergruppen in kompakten zusammenhängenden Liegruppen. *Compositio Mathematica* **38**, No. 2, 129 - 153 (1979)

65. Kung, J. P. S., Rota, G.- C.: The invariant theory of binary forms. *Bulletin Amer. Math. Soc.*, New Ser. **10**, No. 1, 27 - 85 (1984)

66. Lang, S.: *Introduction to algebraic geometry*. New York: Interscience Publishers, Inc. 1958

67. Lang, S.: *Algebra*. Third edition. Reading, Massachusetts: Addison-Wesley Publishing Company 1993

68. Littlewood, D. E.: On invariant theory under restricted groups. *Phil. Trans.*, A, **239**, 387 - 417 (1944)

69. Luna, D.: Adhérences d'orbite et invariants. *Invent. math.* **29**, 231 - 238 ((1975)

70. Luna, D., Vust, Th.: Plongements d'espaces homogènes. *Comment. Math. Helv.* **58**, 186 - 245 (1983)

71. Matsumura, H.: *Commutative Ring Theory*. Cambridge Studies in Advanced Math. **8** Cambridge: Cambridge University Press 1986

72. Mitschi, C., Singer, M. F.: Connected linear groups as differential Galois groups. *J. of Algebra* **184**, no. 1, 333 - 361 (1996)

73. Mumford, D.: *Red Book of Varieties and Schemes*. Lecture Notes in Mathematics **1358**. Berlin - Heidelberg - New York: Springer 1988

74. Mumford, D., Fogarty, J.: *Geometric Invariant Theory*. Third enlarged edition. Ergebnisse der Mathematik und ihrer Grenzgebiete **34**. Berlin - Heidelberg - New York: Springer 1994

75. Nagata, M.: On the fourteenth problem of Hilbert. *Amer. J. Math.* **81**, 766 - 772 (1959)

76. Nagata, M.: *Lectures on the Fourteenth Problem of Hilbert*. Bombay: Tata Institute 1965

77. Newstead, P.E.: *Introduction to Moduli Problems and Orbit Spaces*. Bombay: Tata Institute 1978

78. Panyushev, D. I.: Complexity and rank of homogeneous spaces. *Geom. Dedic.* **34**, 249 - 269 (1990)

79. Panyushev, D. I.: Complexity of quasiaffine homogeneous varieties, t-decompositions, and affine homogeneous spaces of complexity 1. *Advances in Soviet*

Math. **8**, 151 - 166 (1992)

80. Panyushev, D. I.: Some remarks on the deformation method in invariant theory. to appear (1997)

81. Papageorgiou, Y.: $SL_2(C)$: the cubic and the quartic. Ph. D. dissertation, Yale University 1996.

82. Pommerening, K.: Observable radizelle Untergruppen von halbeinfachen algebraischen Gruppen. *Math. Z.* **165**, 243 - 250 (1979)

83. Pommerening, K.: Invarianten unipotenter Gruppen. *Math. Z.* **176**, 359 - 374 (1981)

84. Pommerening, K.: Ordered sets with the standardizing property and straightening laws for algebras of invariants. *Advances in Math.* **63**, 271 - 290 (1987)

85. Pommerening, K.: Invariants of unipotent groups (a survey). In: *Invariant Theory*, 8 - 17. Lecture Notes in Mathematics **1278**. Berlin - Heidelberg - New York: Springer 1987

86. Popov, V.L.: Picard groups of homogeneous spaces of linear algebraic groups and one - dimensional vector bundles. *Math. USSR Izvestja* **8** (2), 301 - 327 (1974)

87. Popov, V. L.: Hilbert's theorem on invariants. *Dokl. Akad. Nauk. SSSR* **249** (1979); Soviet Math. Dokl. **20**, 1318 - 1322 (1979)

88. Popov, V. L.: Modern developments in invariant theory. In: *Proceedings of the International Congress of Mathematicians*, 394 - 406. Berkeley 1986

89. Popov, V.L.: Contractions of the actions of reductive algebraic groups. *Math. USSR Sb.* **58** (2), 311 - 335 (1987)

90. Procesi, C.: *A Primer of Invariant Theory*. Waltham, Mass.: Brandeis Lecture Notes **1** 1982

91. Ramanan, S., Ramanathan, A.: Some remarks on the instability flag. *Tohoku Math. J.* **36**, 269 - 291 (1984)

92. Richardson, R. W.: Affine coset spaces of reductive algebraic groups. *Bull. London Math. Soc.* **9**, 38 - 41 (1977)

93. Roberts, P.: An infinitely generated symbolic blow - up in a power series ring and a new counterexample to Hilbert's fourteenth problem. *J. Algebra* **132**, 461 - 473 (1990)

94. Rosenlicht, M.: Some basic theorems on algebraic groups. *Amer. J. Math.* **78**, 401 - 443 (1956)

95. Rosenlicht, M.: On quotient varieties and the affine embedding of certain

homogeneous spaces. *Trans. Amer. Math. Soc.* **101**, No.2, 211 - 223 (1961)

96. Rosenlicht, M.: A remark on quotient spaces. *Annaes Academia Brasileira de Ciencias* **35**, 487 - 489 (1963)

97. Sampson, J. H., Washnitzer, G.: A Künneth formula for coherent algebraic sheaves. *Illinois J. Math.* **3**, 389 - 402 (1959)

98. Seshadri, C. S.: On a theorem of Weitzenböck in invariant theory. *J. Math. Kyoto Univ.* **1**, 403 - 409 (1962)

99. Seshadri, C. S.: Some results on the quotient space by an algebraic group of automorphisms. *Math. Ann.* **149**, 286 - 301 (1963)

100. Springer, T. A.: *Invariant Theory*. Lecture Notes in Mathematics **585**. Berlin - Heidelberg - New York: Springer 1977

101. Springer, T. A.: Aktionen reduktiver Gruppen. In: *Algebraische Transformationsgruppen und Invariantentheorie*. Basel - Boston - Berlin: Birkhäuser Verlag 1989

102. Steinberg, R.: *Lectures on Chevalley Groups*. Mimeographed lecture notes. New Haven: Yale University Math. Dept. 1968

103. Steinberg, R.: *Conjugacy Classes in Algebraic Groups*. Lecture Notes in Mathematics **366**. Berlin - Heidelberg - New York: Springer 1974

104. Steinberg, R.: Nagata's example. In: *Algebraic Groups and Lie Groups*. Australian Math. Soc. Lecture Series **9**, 375 - 384. Cambridge: Cambridge University Press 1997. Reprinted in *Robert Steinberg: Collected Papers*, 569 - 578. Providence, R.I.: American Math. Society 1997

105. Sturmfels, B.: *Algorithms in Invariant Theory*. Texts and Monographs in Symbolic Computation. Berlin - Heidelberg - New York: Springer 1993

106. Sukhanov, A. A.: Description of the observable subgroups of linear algebraic groups. *Math. USSR Sbornik* **65**, No. 1, 97 - 108 (1990)

107. Tan, L.: The invariant theory of unipotent subgroups of reductive algebraic groups. Ph. D. dissertation, UCLA, Los Angeles 1986

108. Tan, L.: On the Popov - Pommerening conjecture for groups of type G_2. *Algebras, Groups and Geometries* **5**, 421 - 432 (1988)

109. Tan, L.: On the Popov - Pommerening conjecture for groups of type A_n. *Proc. Amer. Math. Soc.* **106**, 611 - 616 (1989)

110. Turnbull, H. W.: *The Theory of Determinants, Matrices, and Invariants*. 3rd ed. New York: Dover Publications Inc. 1960

111. Vinberg, E. B.: Complexity of actions of reductive groups. *Funct. Anal. Appl.* **20**, 1 - 11 (1986)

112. Vinberg, B., Popov, V. L.: On a class of quasi - homogeneous affine varieties. *Math. USSR Izv.* **6**, 743 - 758 (1972)

113. Vust, T.: Sur la théorie des invariants des groupes classiques. *Ann. Inst. Fourier Grenoble* **26**, 1 - 31 (1976)

114. Weitzenböck, R.: Über die Invarianten von linearen Gruppen. *Acta Math.* **58**, 231 - 293 (1932)

115. Weyl, H.: *Theory of Invariants*. Outline by Alfred H. Clifford. Princeton: The Institute for Advanced Study. Fall term 1936

116. Weyl, H.: *The Classical Groups*. Second edition. Princeton: Princeton University Press 1946

117. Zariski, O.: Interprétations algébro - géométriques du quatorzième probléme de Hilbert. *Bull. Sci. Math.* **78**, 155 - 168 (1954)

118. Zariski, O., Samuel, P.: *Commutative Algebra*. vol.II. Princeton: Van Nostrand Company, Inc. 1960

Index